Student Study Guide

to accompany

BOTANY

Second Edition

Moore • Clark • Vodopich

Prepared by
Rebecca McBride DiLiddo
University of Illinois at Chicago

Boston, Massachusetts Burr Ridge, Illinois Dubuque, Iowa
Madison, Wisconsin New York, New York San Francisco, California St. Louis, Missouri

WCB/McGraw-Hill

*A Division of The **McGraw-Hill** Companies*

Student Study Guide to accompany
BOTANY

Recycled/acid free paper
This book is printed on recycled, acid-free paper containing 10% postconsumer waste.

2 3 4 5 7 8 9 0 QPD/QPD 9 0 9 8 7

ISBN 0-697-28631-2

www.mhhe.com

INTRODUCTION

For most college students a botany text is their first introduction to the world of plants. Even students who had biology in high school often find plants to be a foreign subject. This study guide has been developed with these realities in mind.

Reading science texts is not like reading history or fiction. Science texts are dense with facts and full of unfamiliar terms. Learning to read a science text is somewhat like learning to read a foreign language. One must learn the vocabulary and the idioms of the discipline before one can begin to think like a scientist in the field. This Study Guide includes a master plan for reading that will help you get the most from your reading of the text.

Each set of questions in the Study Guide has been developed to assist you in organizing and learning the important points of the chapter. Each question type addresses a different set of skills and represents a different level of difficulty. If you can answer all of the questions for each chapter thoroughly, you will have a good grasp of the content and concepts presented in the chapter.

Botany is a rapidly changing and challenging field. The Study Guide will help you develop the skills to master current knowledge and to learn on your own as new information becomes available.

STUDY GUIDE QUESTIONS

Write out the answers to all questions. You may think that you know the answer to a question, when actually you could not express the answer if called upon to do so. Being able to write out answers in full sentences tests your level of understanding. If you cannot write out the answer, you may not be able to summon up the same information or ideas in another context.

If you have a study group, take turns explaining your answers to other students. By explaining the material to others in your own words you reinforce your knowledge and understanding of the material. Teaching concepts to others is a powerful learning tool.

AFTER YOU HAVE READ THE CHAPTER

These questions should be attempted after you have read the chapter through once. They cover the key ideas and basic facts of the chapter. You should be able to answer most of the questions after one reading of the chapter material; however, you may find that you cannot answer some of the questions. This is not unusual. Read the chapter again following the plan set forth under "Reading Master Plan." This time look for answers to questions that you had trouble with after the first read-through. When you have written answers to all the questions, check them against the answers in the Study Guide. If you answer a question incorrectly, look up the answer on the page given in the Study Guide. Make sure you understand why the answer given is correct. If you do not understand why it is correct or do not agree with the Study Guide answer, check with your instructor.

CONSTRUCTING A CONCEPT MAP

The chapters contain many new words and concepts. Being able to recognize related groups of terms and concepts and understand their interrelationships is an important stage in incorporating knowledge. Constructing a concept map can assist in the process. Below are some suggestions on how to get started and a sample map.

A good place to begin is the chapter outline. It generally lists the main concepts and sub-concepts of the chapter. Look at Chapter 17, "Reproductive Morphology." The main topics are Flowers, Reproductive Morphology and Plant Diversity, Fruits, Seeds, and Dispersal of Fruits and Seeds. These could become the main boxes in a concept map. As you study these and the other headings in the chapter, you will begin to see connections between subtopics within each section and between sections. A sample concept map is shown in figure 1.

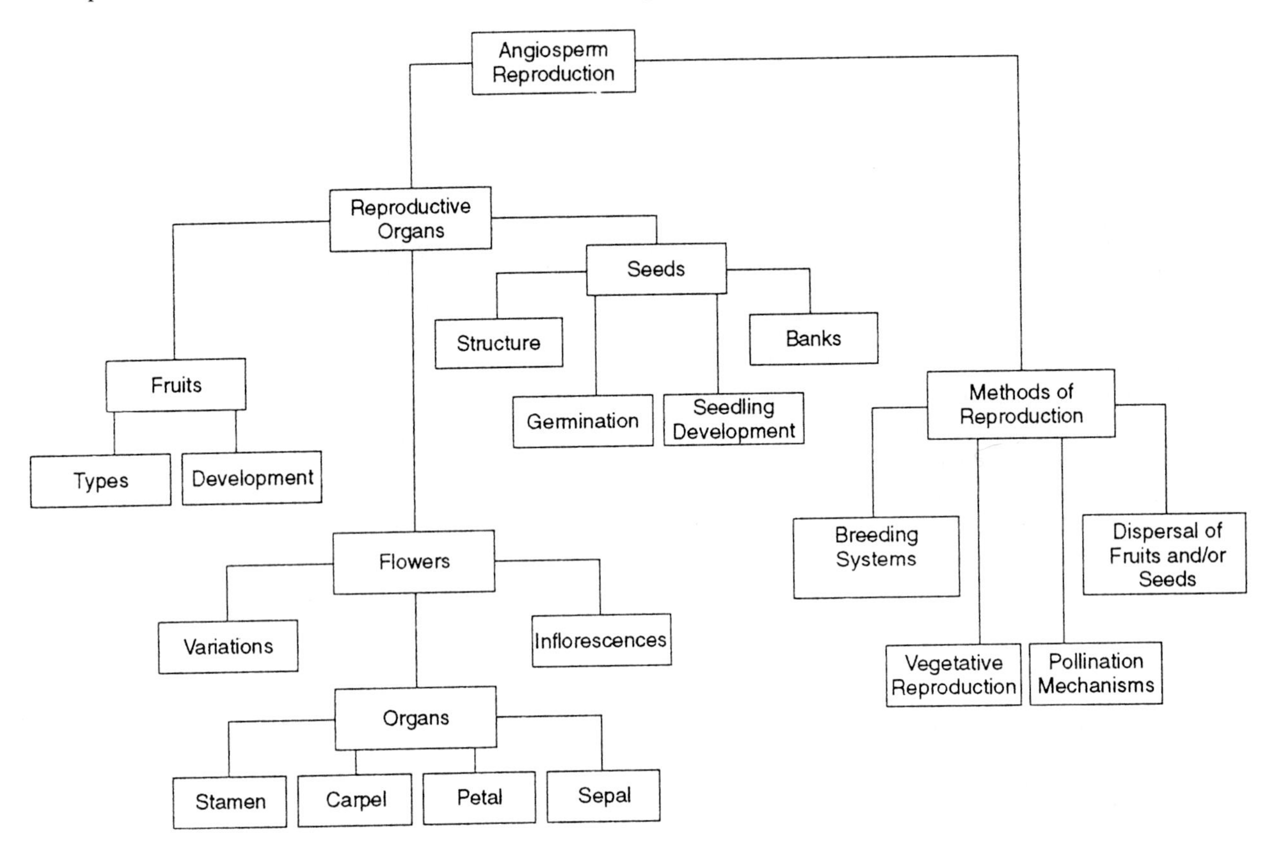

Figure 1 Sample Concept Map

FILL-IN-THE-BLANKS

These questions call for direct recall of facts. The questions allow you to determine if you have mastered the basic concepts and facts of the chapter. The blanks should be filled with terms taken directly from the text. Each question is followed with the page number in the chapter where the answer can be found. If you have trouble with a question, look up the page and reread the appropriate section.

PUTTING YOUR KNOWLEDGE TO WORK

These questions require you to put together more than one concept or idea. They test your mastery of the material beyond the learning of basic facts by asking you to use concepts and facts to figure out problems. Many of the questions present hypothetical situations that simulate real problems faced by botanists as they work. These questions can be difficult. Discuss the answers with classmates and your instructor.

DOING BOTANY YOURSELF

Having a working knowledge of plants that can be used to solve problems and develop new knowledge is more than knowing facts about plants. These questions give you practice in taking facts and applying them to novel situations through the development of experiments. Here you will use the scientific method to design an experiment that involves controls and variables. When you develop the answers to these questions think carefully about the limits of the conclusions that you can draw from the collected data and further experiments that could be done. Clues to solving the problem are given throughout the chapter.

Scientists do not just memorize facts. They must know how to use facts they already know to search for answers to questions that no one knows. Scientists do this by doing experiments. This exercise allows you to practice being a research scientist.

READING MASTER PLAN

Before reading your botany text you need to have a plan. It will increase your comprehension and ability to apply the principles described in the text. The step-by-step plan outlined below will maximize the results of your reading. The steps may seem time-consuming, but if you follow them, you should find that you are actually spending less time studying and getting more out of your reading. As with any new system it may take a little getting used to. But stick with it. The study skills you learn can be applied to any subject.

1. **Examine the organization of the text.** Look at the elements that occur in each chapter and think about what each element tells you.

   Chapter Title. The title tells you the general topic of the chapter.
   Chapter Overview. The overview briefly describes what the chapter will discuss. If you are totally unfamiliar with the subject of the chapter, it will orient you and prepare you to read further.
   Chapter Outline. The outline lists the main headings and subheadings in the chapter. It shows the organization of the material and the approach used by the author. Read each heading in the outline before you read the chapter. This will give you a framework for thinking about the content of the chapter.
   Chapter Summary. The chapter summary can be used to check to see if you have missed any important points. If there are any terms or topics mentioned in the summary that you do not remember clearly from reading the chapter, jot them down. Use your notes to find and reread relevant sections in the chapter.

2. **Read the Chapter Overview.** Jot down the key topics mentioned in the overview. These are points the author will emphasize.
3. **Leaf through the chapter to look at the headings.** Do any of them match the key topics from the overview? If so, you now know at about what point each will appear in the text and can anticipate the sequence of topics. This gives you a mental outline into which you can place information as it is presented in the text.
4. **Read through the chapter fairly quickly.** Do not stop to look up terms you do not know or to take notes. This first reading is to familiarize yourself with the general content of the material and to give you an idea of which material is familiar and which is new.
5. **Answer the "After You Have Read the Chapter" questions in the Study Guide.** Check your answers against those in the Study Guide, noting any that are incorrect or incomplete.
6. **Reread the text more slowly.** This time underline important points and new terms with their definitions. Write in the margins questions that arise as you read. Also note important concepts in the margins. If your margin questions have not been answered by the end of the chapter, use the index to search for answers elsewhere in the text. If you still do not find the answer, ask your instructor or discuss your question with a classmate. If tutoring is available at your institution, these questions can provide a place to begin your session with the tutor.

As you read the text the second time, look for the answers to the questions in the "After You Have Read the Chapter" section. If you missed any answers earlier, check them using the page given with the answer. Be sure you not only know the answer to the question but understand why the answer given is the correct one. An exam may test the same term or concept in a different way. Knowing why an answer is correct will help you answer different questions that require similar responses.

7. **Make a concept map.** Refer to the model in figure 1. Constructing a concept map will help you to see relationships between terms and concepts and incorporate them into your mental outline of the chapter.
8. **Answer the "Fill-in-the Blank" questions.** Each question focuses on a single term or concept. After completing the exercises above, you should be able to respond directly with the appropriate term and concept. Write out your answers, checking each response for correct spelling. A term that is spelled incorrectly is not a correct answer. For example, *inter*fascicular does not mean the same as *intra*fascicular. Learning the correct spelling of botanical terms practices precision in expression and thought, which are important skills in studying and learning science material.
9. **Complete the last two question sets.** You may need to reread certain sections of the chapter to answer these questions. This is normal and does not mean that you are not understanding the chapter content.

Your Study Guide is used in this master reading plan as a tool to increase your reading comprehension. Following the plan will allow you to master the material in a reasonable amount of time.

You are ready to begin using your botany text as a tool to learn about the role of plants in the natural world and the many ways in which humans use plants as sources of food, medicines, and commercial products. An appreciation for the complexity and fragility of our biosphere is essential to creating policies that will sustain the biomes and their inhabitants over the next century. This appreciation begins with an understanding of the relationship between plants and human beings. You belong to a generation whose attitude to all species will determine the fate of our own. Good luck on your learning adventure.

CHAPTER 1 AN INTRODUCTION TO BOTANY

After You Have Read the Chapter

1. In what ways do humans depend on plants for their basic daily needs?
2. What is the role of plants in sustaining life on earth?
3. Describe the types of work done by botanists.
4. Plants hold many records for the tallest, the oldest, etc. What is one such record and which plant holds it?
5. How do scientist use the scientific method to explain observations?
6. How does an hypothesis differ from a guess?
7. Who is Barbara McClintock?
8. What piece of equipment made proof of the cell theory possible?
9. What are two of the unifying themes of botany?
10. Describe two areas of research in plant biotechnology.

Constructing a Concept Map

Using the explanation provided in the introduction to the Study Guide, construct a concept map of the ideas in the chapter.

Fill-in-the-Blanks

1. Life on earth depends on ___________ and _______________ generated by plants. (p. 4)
2. Over __________% of our food supply comes from six species of plants. (p. 4)
3. The tallest living organisms are ___________________, which are found along the California coast. (p. 4)
4. The Union army used the juice from ____________ to cleanse wounds. (p. 8)
5. Spruce trees are used to make ________________ and _______________. (p. 8)
6. The ______________________________ is a process used by botanists and other scientists to find out how the universe works. (p. 9)
7. A ______________________ is a possible explanation for the cause of an observed event that can be tested by experimentation. (p. 9)
8. Barbara McClintock found that ______________________ can move from one position on a corn chromosome to another or even from one chromosome to another. (p.10)
9. Hypotheses can be tested by performing _________________ and observing the results. (p. 10)
10. In part the cell theory states that all living cells must come from ______________________________. (p. 12)
11. The first _____________________ were used to examine cork and the organisms in pond water(p. 12)
12. Plant parts such as leaves and roots look different primarily because of differences in the arrangements of ______________________ present in all plant organs. (p. 14)
13. The conversion of energy in plants occurs by a set of chemical reactions called ______________________. (p. 14)
14 Most plants can reproduce both sexually and ________________________. (p. 15)
15. Charles Darwin outlined his theory of ______________________ in his 1859 book *The Origin of Species.* (p. 15)
16. The first cells probably appeared on earth about__________________ million years ago. (p. 15)

17. The most diverse plant group on earth today is the ________________________. (p. 15)
18. The use of organisms to make commercial products is called ________________. (p. 15)
19. The environmental movement was launched by the publication of the book ____________________, written by Rachel Carson in 1958. (p. 17)
20. Botany tries to instill an appreciation for ______________, which will help stem the effects of humans' longtime neglect of the environment. (p. 17)

Putting Your Knowledge to Work

1. You are a research scientist with a major paper manufacturing company in the United States. You are responsible for finding and developing new sources of fiber for paper. Which of the following would not be a good source to explore?
   A. hemp
   B. elm wood
   C. flax
   D. sisal

2. Rain forests are being cut at a rate that could lead to their total destruction by the end of the century. Which of the following may increase if the rain forests are cut?
   A. the concentration of oxygen in the atmosphere
   B. the concentration of carbon dioxide in the atmosphere
   C. the number of living plant species
   D. the depth of the ozone layer

3. A plant pathologist wants to develop a rose that is resistant to being eaten by caterpillars. She wants to do this by inserting genes for toxic proteins into the roses. When the caterpillar eats the rose leaves, it will be killed by the toxic protein. Which of the following would be a potential source for such a gene?
   A. *Bacillus thuringiensis,* a soil bacterium
   B. *Ceratocsytis ulmi,* a plant fungal pathogen
   C. *Sequoia sempervirens,* a conifer
   D. *Papaver somniferum,* opium poppy

4. An archaeologist wants to determine the climate at the time of the building of the Parthenon in Greece. Climate information can be obtained from analysis of the ring patterns in core samples taken from tree trunks. What would be a possible source of such a sample?
   A. bristlecone pines
   B. giant sequoias
   C. rain forest trees
   D. A or B would have specimens old enough for the core sampling.

5. Which sequence would reflect the steps a scientist would follow when using the scientific method to solve a problem?
   A. observation, experimentation, conculsions
   B. conclusions, hypothesis, observation
   C. experimentation, conclusions, observation
   D. randomization of test subjects, conclusions, hypothesis

Doing Botany Yourself

Aaron Ball is an undergraduate at The Univeristy of Arizona working on a research project in the laboratory of Judy Verbeke.

Aaron describes his research:

My research began with a curious "mutant" plant that was discovered before I began working in the lab. The mutant was found in a seed lot of normal *Cantharanthus roseus*. It is a variety that has no petals. The goal of my project is to determine if the mutant is genetically stable. To do this I must show that the mutant plants can produce offspring. The challenge is that since the mutants have no male parts, cross pollination with pollen from normal plants is necessary.

A basic requirement for pollen to set seed in a receiving plant is that the pollen be able to germinate on the stigma of the receiving plant. Being able to screen pollen/stigma matches would speed up the testing of potential crosses.

Develop a screening technique to determine potential pollen/stigma matches for potential crosses in a petri dish.

Answer Key to Chapter 1 Study Guide Questions

After You Have Read the Chapter

1. Plants provide food, drinks, medicine, clothing, furniture, and many other products. (p. 4)
2. Plants generate oxygen and sugars needed by living organisms. (p. 4)
3. Botanists study plants in many ways; they may study uses of plants by humans, the evolution, structure, and function of plants, the causes of plant extinction, and plant diseases, including their prevention and cure. (p. 1-18)
4. The coastal redwoods, *Sequoia sempervirens,* are the tallest and some bristlecone pines are the oldest. (p. 4)
5. Scientists use the scientific method to develop explanations for events they observe and to develop theories that can predict the outcome of events. (p. 9)
6. A hypothesis is a possible explanation for an observed phenomenon based on known information of how the universe works. A guess provides an explanation not based on observation or knowledge of the event but on personal opinion. (p. 9)
7. Barbara McClintock is a geneticist whose work on ``jumping genes" in the 1940s won her the Noble Prize in 1983. (p. 10)
8. The microscope finally provided proof that all organisms are composed of cells. (p. 12)
9. The unifying themes of botany are these: plants consist of organized parts, plants exchange energy with the environment, plants respond and adapt to their environment, plants reproduce, and plants share a common ancestry. (p. 14)
10. Plant biotechnology is being used to create plants resistant to disease, insects, and herbicides, develop better foods from plant products, find sources of vitamins and medicines, and in many other applications. (p. 15)

Constructing a Concept Map

Your map should incorporate headings from the chapter and show the relationships between main concepts and subconcepts.

Fill-in-the-Blanks

1. oxygen; sugars
2. 80
3. giant sequoias
4. onions
5. newsprint (paper), violins
6. scientific method
7. hypothesis
8. genes
9. experiments
10. other cells
11. microscopes
12. tissues
13. metabolism
14. asexually
15. evolution
16. 3.6 billion
17. flowering plants
18. biotechnology
19. *Silent Spring*
20. living organisms

Putting Your Knowledge to Work

1. B. elm wood
2. A. the concentration of oxygen in the atmosphere
3. A. *Bacillus thuringiensis*
4. D. A (bristlecone pines) or B (giant sequoias) would have specimens old enough for necore sampling.
5. A. observation, experimentation, conclusions

Doing Botany Yourself

The follownig could be used as a screening technique:

1. Using stigmas from the mutants make a water extract. Apply the extract to pollen samples from potential receiving plants that is held in the well of a depression slide.
2. Apply a coverslip.
3. Count the number of pollen grains in the sample.
4. Wait 1hr, 3hrs, 6hrs, etc.
5. Count the number of pollen grains from the original sample that germinate at each time interval.
6. The sources of stigma extracts that support the highest percentage of pollen germination are the best potential receivers for the mutant pollen.

CHAPTER 2 ATOMS AND MOLECULES: THE BUILDING BLOCKS OF LIFE

After You Have Read the Chapter

1. What are the four elements that compose more than 99% of living organisms?
2. Give three examples of biological polymers.
3. In what ways do humans exploit the glue-like properties of many polysaccharides?
4. What are the two most abundant forms of polysaccharides in plants?
5. Describe the types of biopolymers found in the cell walls of plants.
6. What are three roles of proteins in plants?
7. What are the functions of nucleic acids in plants?
8. Compare DNA to RNA.
9. Differentiate the three major types of lipids found in plants.
10. What are the three major types of secondary metabolites and what do they do in plants?

Constructing a Concept Map

Using the explanation provided in the introduction to your Study Guide construct a concept map of the ideas in the chapter.

Fill-in-the-Blanks

1. Most molecules that contain carbon are considered to be ___________ compounds. (p. 22)
2. ___________ are organic and inorganic compounds that occur in living organisms. (p. 22)
3. Sugar, cellulose, and starch are all ________________________. (p. 22-25)
4. Amylose and cellulose are polymers of _____________. (p. 25)
5. Plants store carbohydrates as _________________. (p. 26)
6. Only organisms that can produce ___________________ can use cellulose as a source of glucose. (p. 27)
7. The amino acids in a polypeptide are joined by ___________ bonds. (p. 27)
8. Proteins are polymers of ________________________. (p. 27)
9. Seeds contain high concentrations of storage ___________ that serve as sources of amino acids for developing seedlings. (p. 30)
10. Many cereal grains contain _________________, that inhibit the digestive enzymes of animals. (p. 30)
11. _____________ are proteins that catalyze biochemical reactions. (p. 30)
12. Chemicals and heat can ____________________ proteins disrupting their structure and preventing them from functioning properly.(p. 28)
13. Nucleic acids are polymers of ______________________. (p. 31)
14. In RNA _________________ replaces thymine. (p. 31)
15. The number of adenines in a molecule of DNA is equal to the number of ________________ molecules. (p. 32)
16. Oils are _______________ that are liquid at room temperature. (p. 33)
17. Tropical oils are examples of _______________ fats produced by plants. (p. 34)
18. Cuticular wax and _________________ make plant surfaces water-repellent. (p. 35)

19. Secondary compounds like ____________________ can cause dramatic physiological effects on humans. (p. 36)
20. Phenolic compounds include salicylic acid in willows and the polymer ______________ , which serves as a strengthening agent in woody tissue. (pp. 39)

Putting Your Knowledge to Work

1. You and your friends are on a camping trip. You catch several trout. After cleaning the fish, you start a fire. A rainstorm drives you into the tents with your raw "dinner". A quick check of the provisions turns up three lemons. Your dinner is saved. Why?
   A. Lemon makes a good condiment for any fish dish.
   B. The acid in the lemon will kill any bacteria on the fish so it can be eaten raw like sushi.
   C. Soaking the fish in lemon juice will denature or "cook" its protein the same way as heat does.
   D. Drinking lemonade will take away the "fishy" taste of the raw fish.

2. A new diet pill promises to make you feel full after even a small meal so that you can lose weight and never feel hungry. It claims to contain no chemicals, only "natural" ingredients. As you read the label you are most likely to find that the main ingredient in the tablets is:
   A. fibrous cellulose.
   B. starch.
   C. hydrolyzed protein.
   D. sucrose.

3. You are an apprentice chef at a four-star hotel in Hawaii. One day you notice that the head pastry chef is about to stir fresh pineapple juice into a gelatine-based mousse. You stop her. What do you know about pineapple juice that she doesn't?
   A. Fresh pineapple juice contains a toxic alkaloid.
   B. Papain in the fresh pineapple juice will prevent the gelatin in the mousse from becoming firm.
   C. Only frozen pineapple juice should be used in gelatin deserts.
   D. An enzyme in fresh pineapple juice polymerizes proteins making the mousse too stiff.

4. You are a biochemist. You have isolated a large nucleic acid molecule from a cell extract. The molecule is a single helix with a content of 40% uracil. Which of the following could you say with confidence about this molecule?
   A. The molecule is an RNA molecule.
   B. The thymine content of the molecule is 40%.
   C. The guanine content of the molecule is 20%.
   D. The molecule will occur at its highest concentrations in the nucleus.

5. A new stick-margarine sold by a chain of health food stores claims to be a healthy alternative to the saturated fat in butter because it is made from 100% corn oil. This is a misleading claim because
   A. corn oil is similar in fat content to tropical oils.
   B. oils are all saturated fats.
   C. corn oil is a lipid, not a fat.
   D. corn oil must be converted to a saturated fat in order to be formed into stick

Doing Botany Yourself

A reseracher at Nutrasweet receives several kilograms of a water extract produce from the leaves of a tropical tree. She is thinking about developing the sweetener as a sugar substitute. To do so she must design an experiment to determine its sweetening ability. She gives the assignment for designing the experiment to you her assistant. What did you suggest?

Answer Key to Chapter 2 Study Guide Questions

After You Have Read the Chapter

1. The four elements are hydrogen, carbon, oxygen, and nitrogen. (p. 22)
2. Biopolymers include cellulose, starches, enzymes, DNA, waxes, and lignin. (p. 22-37)
3. Gum arabic is used to stabilize postage stamp glue, beer suds, hand lotions, and liquid soaps; agar from algae are used in drug capsules, cosmetics, gelatin deserts, and a medium for growing; carrageenan from algae is used as a stabilizer in paints, cosmetics, salad dressings, and dairy products. (p. 22-37)
4. The most abundant polysaccharides are cellulose and starch. (p. 24-27)
5. Cell walls contain cellulose microfibrils(glucose polymer); pectins (polymers of galact-uronic acid); hemicelluloses(polymers of xylose); extensins(glycoproteins high in hydroxyproline). (p. 25)
6. Proteins are structural, storage, and enzyme molecules. (p. 25,27)
7. DNA carries hereditary information and RNA directs protein synthesis. (pp. 31)
8. DNA is a double-stranded a helix with one strand carrying the genetic code in the sequence of nucleotides. Each adenine is paired to a thymine and each cytosine to a guanine to form the genetic code. DNA contains deoxyribose.DNA is found mostly in the nucleus in the chromosomes. RNA is a single-stranded molecule that plays three roles in protein synthesis. Each adenine is paired with a uracil and each cytosine with a guanine. RNA contains ribose sugar. RNA is made in the nucleus and moved to the cytoplasm.(p. 31-32)
9. Lipids include oils(fats composed of fatty acids and glycerol), phospholipids(membrane fats with a phosphate replacing a fatty acid making them water soluble at one end and fat soluble at the other), tropical oils(saturated plant fats),waxes(lipids composed of fatty acids and long-chain alcohols that are harder and more water-repellent than other lipids); and wax-like (cutin and suberin composed of hydroxylated fatty acids). (p. 32-36)
10. Secondary metabolites include alkaloids, terpenoids, and phenolics. These compounds govern the interactions between plants and other organisms. (pp. 36-41)

Constructing a Concept Map

Your map should incorporate headings from the chapter and show the relationships between main concepts and subconcepts.

Fill-in-the-Blanks

1. organic
2. Biochemicals
3. carbohydrates
4. glucose
5. starch
6. cellulase
7. peptide
8. amino acids
9. protein
10. protease inhibitors
11. Enzymes
12. denature
13. nucleotides
14. uracil
15. thymine
16. lipids or fats
17. saturated
18. cutin
19. alkaloids
20. lignin

Putting Your Knowledge to Work

1. C. Soaking the fish in lemon juice will denature or "cook" its protein the same way as heat does.
2. A. fibrous cellulose
3. B. Papain in the fresh pineapple juice will prevent the gelatin in the mouse from getting firm.
4. A. The molecule is an RNA molecule.
5. D. corn oil must be hydrogenated in order to be formed into sticks

Doing Botany Yourself

A sweetener is used to replace table sugar or sucrose. Comparing the sweetener to sucrose will tell the researcher how much of the sweetener will have to be added to a food to make it as sweet tasting as when sugar is added. The amount and its cost will determine if the sweetener will make a good sugar replacement.

Providing the sweetener in a liquid form will be convenient and will take the least preparation time.

You suggest that two series of drinks be prepared one using known amounts of sugar and another using known amounts of the sweetener. The subjects will not know which of the drinks contains sugar or sweetener. A sugared drink is tasted by a subject. Then a drink made with sweetener is tasted with the same amount of sweetener as sugar. The subject will tell you if the second drink is sweeter or less sweet than the sugared drink. Tests can reverse the order of the drinks.

By comparing the responses of a number of subjects to drinks containing known amounts of sugar you will be able to determine how much more or how much less sweetener to add to a drink to get the same sweetness as with sugar.

CHAPTER 3 STRUCTURE AND FUNCTION OF PLANT CELLS

After You Have Read the Chapter

1. Compare the properties of magnification and resolution of light microscopes to that of electron microscopes.
2. Describe three ways that cytologists study cells.
3. What factors limit cell size?
4. Describe the types of structures that make up the cytoskeleton.
5. What are two functions of cellular membranes?
6. Compare the primary cell wall to the secondary cell wall.
7. Explain how cells grow.
8. What are two functions of the central vacuole of a plant cell?
9. Describe the roles the nucleus plays in protein synthesis.
10. How does the presence of plasmodesmata between most plant cells lend support to the organismal theory of plant organization?

Constructing a Concept Map

Using the explanation provided in the introduction to your Study Guide, construct a concept map of the ideas in the chapter.

Fill-in-the-Blanks

1. The ability of a microscope to distinguish two objects as individual bodies refers to the ____________ ____________ of the microscope. (p. 46)
2. In an electron microscope a beam of ____________ is used to ``see" the specimen. (p. 46)
3. A ________________________ studies cell structure and function. (p. 48)
4. The ____________________________________ is used to obtain three-dimensional pictures of cells and other small objects. (p. 50)
5. _______________________ is the process of separating cell organelles in a cell suspension on the basis of differences in density. (p. 52)
6. The _________________________ regulates entry and exit from the cell. (p. 54)
7. The pectins of the _________________ form a glue that holds to adjacent plant cells together.(p. 56)
8. The ______________________ separates the contents of the nucleus from the cytosol. (p..58)
9. Protein and lipid synthesis is carried out in the _______________________ formed from a series of membrane tubes throughout the cytosol. (p. 51)
10. The ____________________ in the cell regulates turgor pressure by taking up and releasing water.(p. 57)
11. Both peroxisomes and glyoxysomes are considered to be ______________________ because they contain catalases to detoxify hydrogen peroxide.(p. 62)
12. DNA complexes with proteins in the nucleus to form ______________________ (p. 58)
13. Complex molecules made in the ER are processed by the ______________________ before release by vesicle fusion at the plasma membrane. (p. 61)
14. Dictyosome vesicles containing cell wall material fuse to form the________________ during cytokinesis in a dividing plant cell.(p. 61)

15. Photosynthesis takes place in plant cells in the chlorophyll-containing plastids called ____________________________. (p. 63-64)
16. Organic materials produced by photosynthesis are converted to ATP for cell metabolism in the ________________________. (p. 64)
17. The process of ________________________ moves chloroplasts and mitochondria, enhances exchange of materials between organelles, and mixes the contents of the cell. (p. 65)
18. Chloroplasts and mitochondria probably originated from ________________ taken in by a host cell. (p. 66)
19. All ________________ in eukaryotic organisms have an internal structure of a ring of nine microtubules surrounding two central microtubules. (p. 66)
20. The ____________________ theory of biological organization views the organism as a large cell and the cells as compartments. (p. 68)

Putting Your Knowledge to Work

1. A cytologist must decide what type of microscope to purchase for his laboratory. The lab examiines fibers, hairs, and other forensic evidence as whole specimens. As the microscope salesperson you suggest that he should buy:
   A. a light microscope with resolution power to 1nm.
   B. a scanning electron microscope.
   C. a transmission electron microscope with maximum magnification of 1,000X.
   D. an electron microscope with resolution power of about 2 nanometers.

2. You have developed a stain that colors alpha and beta tubulin a bright orange color. When the stain is applied to non-dividing leaf cells, what area of the cell would you expect to be orange when viewed with a light microscope?
   A. the nucleus
   B. the cell wall adjacent to the surface of the plasma membrane
   C. the cytoskeleton
   D. flagella

3. A lab technician spun a cell suspension from leaf tissue in a centrifuge in a slurry of sucrose at a speed that forms a bright green band about half way down the solution in the tube? In what fraction would you expect to find the nuclei?
   A. in a pellet at the bottom of the tube
   B. in the green band
   C. in a dark band above the green band
   D. in the supernatant

4. Mitochondria and chloroplasts are thought to have evolved from bacteria taken in by a host cell. If this is true, which of the following would a cytologist expect to find?
   A. a cyanobacterium with mitochondria
   B. a green alga with chloroplasts, but no mitochondria
   C. all chloroplasts have developed from one source
   D. amoebae that take in cyanobacteria when food is scarce and use them as chloroplasts

Doing Botany Yourself

Devise an experiment to determine if the arrangement of chloroplasts in plant cells is the same in all plants.

ANSWER KEY TO CHAPTER 3 OF THE STUDY GUIDE

After You Have Read the Chapter

1. The limit of magnification of a light microscope is about 1000X's; an electron microscope about 100,000X's. the resolving power of a light microscope is about 2 micrometers or the size of a mitochondrion. The resolving power of the electron microscope is about 4nms. (p. 46-48)
2. A cytologist uses the light microscope, electron microscopes, and cell chemistry. (p. 48-50)
3. Factors that regulate cell size are surface-to-volume ratio, rates of synthesis and transport inside the cell, and the limit of one nucleus per cell. (p. 53)
4. The cytoskeleton is composed of a variety of filments that interact to create a dynamic scaffolding for thecell. Microtubules composed of alpha and beta tubulin are the largest. The smallest are actin filaments composed of two strands of actin each. Intermediate filaments vary in composition from cell type to cell type. (p. 54-55)
5. The plasma membrane regulates entry and exit of materials from the cell; the tonoplast regulates entry and exit rom the vacuole; internal cell membranes provide surfaces for metabolic processes and form compartments within the cell. (p. 59-62)
6. The primary and secondary cell walls are composed of cellulose, pectins, hemicelluloses, proteins, and lignin. The primary cell wall is the first formed in young cells and contains less than 25% cellulose and sometimes a little lignin. The primary wall is thin and flexible allowing the cell to change shape, divide, or differentiate. The secondary cell wall forms inside the primary cell wall when cells stop growing. Secondary cell walls are rigid with up to 25% liginin content and lack glycoproteins. Suberin may be present in these walls.(p. 55-57)
7. New cell wall material is manufactured in the dictyosomes and moved to the cell wall through exocytosis. Cell expansion requires that cellulose microfibrils separate and the cell stretch along axes determined by the location of microtubules associated with the wall. (p. 56)
8. The central vacuole regulates turgor pressure and is the site of digestive enzyme action, storage of pigments, toxins, and ions. (p. 62)
9. DNA in the chromosomes carries the genetic code in its nucleotide sequence for the synthesis of mRNA. The nucleotide sequence in mRNA corresponds to the amino acid sequence in proteins. The proteins are made on ribosomes containing rRNA made in the nucleus and using tRNA's made in the nucleus. (p. 58)
10. Plasmmodesmata are plasma membrane-lined channels between cells that allow the movement of materials and objects as large as viruses to pass from one cell to another. Direct connection between adjacent cells could argue for the view that a plant is one continuous cell with compartments(cells) that function much as organelles are thought to function. (p. 57).

Constructing a Concept Map

Your map should incorporate headings from the chapter and show the relationships between main concepts and subconcepts.

Fill-in-the-Blanks

1. resolving power
2. electrons
3. cytologist
4. scanning electron microscope
5. Cell fractionation
6. plasma membrane
7. middle lamella
8. nuclear envelope
9. endoplasmic reticulum
10. central vacuole
11. microbodies
12. chromosomes
13. dictyosomes
14. cell plate
15. chloroplasts
16. mitochondria
17. cyclosis
18. bacteria
19. flagella
20. organismal

Putting Your Knowledge to Work

1. B. a scanning electron microscope.
2. C. the cytoskeleton
3. A. in a pellet at the bottom of the tube
4. D. amoebae that take in cyanobacteria when food is scarce and use them as chloroplasts

Doing Botany Yourself

Obtain samples of leaf tissue from a variety of plants adapted to different types of environments(submerged and floating aquatics, desert plants, crop plants, forest floor wildflowers, etc.).

1. Prepare freehand sections of fresh material.

2. Examine the material at various levels as you focus down through the specimen.

3. Make a diagram of the location of the chloroplasts at each level in each specimen.

4. Using this chloroplast "map" place the chloroplasts in the cell in 3-dimensions.

5. To analyse the results compare the distribution of the chloroplasts in the different species. Are they all distributed in the same pattern? Do species adapted to the same habitat have the same distribution? Is there any relationship between habitat and chloroplast distribution?

CHAPTER 4 MEMBRANES AND MEMBRANE TRANSPORT

After You Have Read the Chapter

1. Describe the basic structure of all biological membranes.
2. What are two functions of cell membranes?
3. What factors affect water uptake by plant cells?
4. Explain the terms hypotonic, isotonic, and hypertonic when describing the concentration of solutes in a cell?
5. Explain why plants wilt.
6. Describe three ways that a solute molecule can be moved across a cell membrane.
7. Explain how plants get large amounts of protective substances manufactured in their cells out on the surface of leaves or roots.
8. What role do proton pumps play in plant cells?
9. Compare a living plant cell to a battery.
10. How can plant cells communicate with one another and with the environment?

Constructing a Concept Map

Using the explanation provided in the introduction to your Study Guide, construct a concept map of the ideas in the chapter.

Fill-in-the-Blanks

1. The ________________ model describes cell membranes as phospholipid bilayers with proteins embedded in bilayer. (p.74)
2. Ions pumps use energy derived from the hydrolysis of ______________ to move ions across cell membranes. (p.76)
3. In an aqueous solution of sucrose the solvent is ____________. (p. 76-78)
4. If a cell is hypotonic to an aqueous solution, water will move __________the cell. (p. 77-79)
5. The __________ potential of a cell is generated by the solutes in the cell. (p. 78)
6. Water moves by _________________ across the plasma membrane. (p. 79)
7. The outward pressure of the plasma membrane against the plant cell wall is called ________________________. (p. 80)
8. Loss of turgor in a hypertonic solution results in __________ of the cell. (p. 80)
9. ________________ requires a transport protein and metabolic energy to move solutes against a concentration gradient. (p. 82)
10. The property of ____________________ allows the plasma membrane to regulate what substances enter and leave the cell. (p. 79)
11. Facilitated diffusion and active transport require special __________________ to move solutes across cell membranes. (p. 81)
12. A difference in concentration of ions across a cell membrane produces an __________ potential across the membrane that can be used to transport ions or to make ATP. (p. 83-85)

13. ________________ uses the energy from a hydrogen ion gradient to move uncharged molecules like sugars across cell membranes. (p. 82)
14. The ____________________ is maintained across the plasma membrane by ion pumps. (p. 84)
15. The pH of the cell wall is regulated by ______________. (p. 84)
16. External signals can be transferred into the cell by membrane molecules such as _________, which activate protein kinases. (p. 85-87)
17. Auxin, like many hormones, affects cell activities by binding to protein _____________ on the plasma membrane. (p. 87)
18. Plant bacterial pathogens like *Pseudomonas syringae* secrete protein ______________, which cause the disintegration of the plasma membrane. (p. 87)
19. Plants can recognize pathogens that infect their cells by ____________________, which are secreted by the pathogens. (p. 87)
20. ____________________ are glycoproteins in root-hair cell walls that allow a symbiont such as *Rhizobium* to recognize its specific host. (p. 88)

Putting Your Knowledge to Work

1. A plant physiologist studying transport of sugars across root cell plasma membranes. What does he need to know to determine if the transport is occurring by facilitated diffusion or active transport?
   A. if a specific transport protein is involved
   B. if the sugar moves across faster if the cell is placed in a more concentrated solution
   C. if ATP are generated when the sugar enters the cytoplasm
   D. can the concentration of sugar in the cell get to be higher than outside the cell

2. A student is given a 1,000-ml beaker of water and a small crystal of methylene blue. Her task is to get the methylene blue into solution without stirring within the three-hour lab period. Which of the following would not speed up the dispersal of the methylene blue throughout the solution?
   A. stirring the solution after adding the crystal
   B. pulverizing the crystal before adding it to the beaker
   C. coating the crystal with vegetable oil before adding it to the beaker
   D. heating the water in the beaker to 95°C before adding the crystal

3. A lab technician is given a bag made of dialysis tubing filled with water. Dialysis tubing allows only water and small molecules or ions to pass through. The technician places the bag into a beaker of 0.5 M sucrose solution. After four hours she returns and removes the bag from the beaker. What will she expect to find?
   A. The bag has expanded.
   B. The bag has gained weight.
   C. The bag has lost weight.
   D. The concentration of sugar in the bag and the beaker are equal.

4. When you put the groceries away you left the leaf lettuce on the counter. The next evening when you are preparing the salad for dinner you discover the lettuce and it is wilted. How can you quickly crisp the lettuce for the salad?
   A. Put the lettuce in the freezer for 10 minutes.
   B. Place the lettuce in a pan of warm water.
   C. Place the lettuce in a large basin of ice water.
   D. Put it in the vegetable crisper of the refrigerator for 10 minutes.

5. A botany student has been observing cultures of *Bacillus subtilus* bacteria to determine what types of substances can be used as food by each strain. He finds that one strain grows well on fructose when the medium has a $pH < 6.0$, but it does not grow well on fructose when the medium has a $pH \geq 6.0$. He also finds that the concentration of fructose inside the bacterial cells is never greater than the concentration of fructose in the medium. Using your knowledge of the many transport systems observed in cells, give an explanation of these observations.
   A. Fructose is probably moved across the membrane by active transport.
   B. Fructose is probably moved into the cell by facilitated diffusion.
   C. ATP is probably required for the movement of fructose across the membrane.
   D. Fructose is probably moved by coupled cotransport as H ions are pumped out of the cell.

Doing Botany Yourself

Using only 15ml test tubes, a 25ml graduated cylinder, a 1M solution of sucrose, and a 1000ml beaker of water determine the solute concentration in the cells of an *Elodea* leaf. *Elodea* is a submerged aquatic plant.

Answer Key to Chapter 4 Study Guide Questions

After You Have Read the Chapter

1. All biological membranes contain a bilayer of phospholipids embedded with proteins. (p. 74)
2. Membrane functions include movement of water and solutes, selective permeability, ion pumps, enzyme activity, and cell communication. (p. 74)
3. Water uptake in cells is determined by the difference in water potential outside and inside the cell. Water moves from an area of higher water potential to an area of lower water potential. Water potential is directly related to the solute concentration of solutions and cell contents and the turgor pressure of the cell. (p. 78)
4. The terms apply to solutions separated by a differentially permeable member that lets water, but not solute particles pass through. If a cell is surrounded by a solution that has the same solute concentration as the cell then the cell is isotonic to the solution and the solution is isotonic to the cell. If the solution has a higher solute concentration than the cell then the solution is hypertonic to the cell and the cell is hypotonic to the solution. If the solution has a lower solute concentration than the cell then the solution is hypotonic to the cell and the cell is hypertonic to the solution. Water moves from the hypotonic compartment to the hypertonic compartment. (p. 80)

5. Plants wilt when the cells lose turgor because they are growing in soil in which the solute concentration is greater than that in the root cells. The roots cannot replace the lost water. The loss of turgor causes the plants to wilt. (p. 80)
6. The processes that move solutes across cell membranes are diffusion, facilitated diffusion, active transport (cotransport and coupled cotransport), and ion pumps. (p. 81–85)
7. Large amounts of protective materials such as waxes or polysaccharides are secreted from the cell by exocytosis using dictyosome vesicles. (p. 83)
8. Proton pumps help to maintain the membrane potential across the plasma membrane; provide energy for the couples cotransport of uncharged solutes like sucrose; regulate the pH of the cell, vacuole and other organelles. (p.84-85)
9. The plasmamembrane is able to separate charged ions to generate a membrane potential (voltage) just as a battery separates charges between two compartments. This voltage potential can be used to do work, just as the voltage in a battery can generate electricity to do work. (p. 84)
10. Membrane proteins assist in cellular communication through the binding of plant hormones, microbial toxins; lectins that allow plants to recognize pathogens and symbionts; and glycoproteins that allow for the recognition of compatible pollen and stigmas for reproduction. (p. 85-87)

Constructing a Concept Map

Your map should incorporate headings from the chapter and show the relationships between main concepts and subconcepts.

Fill-in-the-Blanks

1. fluid mosaic
2. ion pumps
3. sucrose
4. into
5. bars or Megapascals
6. osmosis
7. turgor pressure
8. plasmolysis
9. Active transport
10. differential permeability
11. equal to
12. electrochemical
13. coupled cotransport
14. membrane potential
15. H^+-ATPases
16. calmodulin
17. receptors
18. toxins
19. polysaccharides
20. Lectins

Putting Your Knowledge to Work

1. D. can the concentration of sugar in the cell get to be higher than outside the cell?
2. C. coating the crystal with vegetable oil before adding it to the water
3. C. The bag has lost weight.
4. C. Place the lettuce in a large basin of ice water.
5. B. Fructose is probably moved into the cell by facilitated diffusion.

In a plant cell water potential is equal to the osmotic potential, that portion of the water potential related to solute concentration, when there is no turgor pressure in the cell. Turgor pressure becomes zero just as the tissue is plasmolyzed.

If you can determine the concentration of sucrose solution required to just start plasmolysis in the leaf, this will be approximately equal to the concentration of solute particles in the leaf tissue.

1. Make a series of solutions of concentrations 1M, 0.8M, 0.7M, 0.6M, 0.5M, 0.4M, 0.3M, 0.2M, and 0.1M.

2. Place a leaf into each test tube of solution in the series. Allow the leaves to remain in the solution for 30 minutes.

3. Remove each leaf and examine it under the microscope. Find the leaf in which about 50% of the leaf cells are just beginning to plasmolyze. This leaf was in a solution in which the solute concentration is approximately the same as that inside the leaf cells.

CHAPTER 5 ENERGY AND ITS USE BY PLANTS

After You Have Read the Chapter

1. What is energy and how is it measured?
2. Most energy available to life on earth comes from what source?
3. What are the two basic types of energy?
4. Which universal law of physics might be stated, "You can't get something for nothing" ?
5. Using the laws of thermodynamics, explain why you always have to refill your gas tank.
6. Contrast exergonic and endergonic reactions.
7. If energy for life comes from photosynthesis, then why do plant cells have mitochondria?
8. Distinguish between a cofactor, a coenzyme, and a cytochrome.
9. What is the general function of enzymes and how can their activity be regulated in the cell?
10. Compare the pathway of nutrients and energy in an ecosystem.

Constructing a Concept Map

Using the explanation provided in the introduction to your Study Guide, construct a concept map of the ideas in the chapter.

Fill-in-the-Blanks

1. The study of energy transformation in living organisms is called __________. (p. 94)
2. Raising 3 grams of water 3°C would require 3 ________. (p. 95)
3. When we measure the temperature of a solution, we are measuring ____________ energy. (p. 96)
4. Thermodynamics is the study of the transformation of ____________ from one form to another. (p. 96-98)
5. Plants convert the radiant energy of sunlight into __________________ energy stored in the bonds of carbohydrate molecules. (p. 96)
6. Cellular _________________ is the sum of all the catabolic and anabolic reactions in a cell. (p. 99)
7. The ____________________ released when sugar is metabolized can be used to form ATP from ADP and inorganic phosphate.(p. 100)
8. If enthalpy in a system is increasing then _________________ is decreasing. (p. 100)
9. In an exergonic reaction the products store ________ energy than the reactants. (p. 101)
10. Death of a cell occurs when the cell processes reach____________________. (p. 101)
11. When NAD+ gains a proton it is ______________________.
12. ___________ reactions are usually associated with catabolism. (p. 101)
13. Metabolic reactions that synthesize more complex molecules from simpler molecules are called __________________. (p. 101)
14. Energy stored in fats and carbohydrates is used to generate __________ by cellular metabolism. (p. 102-103)
15. The transfer of a phosphate group from ATP to another molecule is called ___________. (p. 103)
16. Ions such as Na^+ and Mg^{2+} that assist in energy transformations in cells are called ______________________. (p. 103)
17. Coenzymes transfer _________ or_________ between biological molecules. (p. 103)
18. Enzymes lower the _______________________ for metabolic reactions. (p. 105)

19. The level of products of enzyme reactions are kept constant by ________________ inhibition.(p. 106-107)
20. Molecules that compete with a substrate for the active site on an enzyme molecule are called ________________________________. (p. 107)

Putting Your Knowledge to Work

1. A mole of glucose has a delta G of –686 kcal per mole. How many ATP molecules could be synthesized from the energy released from a mole of glucose when it is oxidized to carbon dioxide and water during cellular respiration?
   A. 96.7
   B. 5,007.8
   C. less than 48
   D. None. All the energy is lost as heat.

2. A person is going on a diet. He wants to be able to eat the most food bulk, to keep from feeling hungry, with the least amount of calories. Which of the following would be a good choice?
   A. steak
   B. salads of lettuce and tomatoes
   C. french fries
   D. peanut butter sandwiches

3. A biochemist is studying a new antibiotic extracted from the bark of a tropical tree that seems to inhibit protein synthesis in bacteria. She notes that when she increases the concentration of amino acids in her experimental media, that the inhibition of protein synthesis by the antibiotic is reduced. This tells her that:
   A. the antibiotic is a competitive inhibitor.
   B. the effects of the antibiotic will be reduced in patients on a high protein diet.
   C. the antibiotic is a noncompetitive inhibitor.
   D. the antibiotic is a nonallosteric inhibitor.

4. An ecologist is concerned that we are running out of energy to support life on earth. He wants to estimate how much energy is potentially available to living organisms on earth. What figure should he calculate?
   A. the number of calories in the world's plant biomass
   B. the amount of kinetic energy in the hydrogen
   C. the sun's energy that reaches the earth's surface
   D. the amount of energy released from the burning of fossil fuels.

5. An elderly man has a limited food budget. He needs 2,000 calories per day to maintain his activity level and not lose weight. Currently, he buys and consumes 2,000 calories of food, but he is losing weight. What did he forget?
   A. the law of bioenergetics
   B. the second law of thermodynamics
   C. the first law of thermodynamics
   D. the law of potential energy

Doing Botany Yourself

You are a food lab technician. You obtained the data below on the protein-degrading enzyme in pineapple. A client wants to know whether pineapple chunks to be used in making gelatin (a protein-based food) should be fresh, frozen, or canned. What would you tell your client based on the data below from your experiments and why?

Temperature	**Product Produced in 1 Hour**
0°C	1/4 of gelatin dissolved
10°C	1/2 of gelatin dissolved
20°C	2/3 of gelatin dissolved
25°C	all gelatin dissolved
30°C	1/2 of gelatin dissolved
35°C	1/10 of gelatin dissolved
50°C	no gelatin dissolved
100°C	no gelatin dissolved

Answer Key to Chapter 5 Study Guide Questions

After You Have Read the Chapter

1. Energy is the ability to do work. Energy is measured in calories, Calories, and joules. A calorie is the amount of energy required to raise the temperature of a gram of water one degree. A Calorie is equal to 1000 calories and is the unit normally used to measure energy. A joule is the amount of energy needed to move 1 kilogram through 1 meter with an acceleration of 1 meter per second. 1 calorie=4.12 J. (p. 94-95)
2. The sun is the major source of energy for life on earth. Most organisms rely directly or indirectly on photosynthesis powered by the sun for their organic compounds that are metabolized to release energy. The sun also powers the hydrologic cycle and generates the winds that distribute energy and influences climate around the globe. (p. 96-97)
3. Two basic types of energy are potential energy and kinetic energy. (p. 96)
4. The expression paraphrases the first law of thermodynamics, or the law of conservation of matter and energy. (p. 97-98)
5. The chemical energy in the gasoline is converted to forms of energy that are used to run the parts of the engine and move the car. Every energy conversion results in some energy loss. The energy in the gasoline is used and must be replaced by energy from more gasoline. (p. 97-98)
6. Exergonic reactions release energy, produce products with less free energy than the reactants, are spontaneous, and increase entropy. In endergonic reactions energy is absorbed, and the free energy of the products is more than the substrates. Endergonic reactions are not spontaneous and may decrease entropy. (p. 100-101)
7. Photosynthesis stores the sun's energy in the chemical bonds of sugar molecules. To be used for cell metabolism the bond energy must be converted to ATP. This is accomplished by through cellular respiration. In aerobic organisms the mitochondria are the sites of most ATP production. (p.102-103)
8. Energy transformations reactions in cells often require enzymes and usually involve the transfer of protons and/or electrons between molecules. Cofactors are nonprotein helpers such as ions that are required for the reactions. Coenzymes are organic compounds(vitamins) that can carry protons or electrons such as nicotinamide adenine dinucleotide. Cytochromes are molecules that contain metal ions that participate as electron carriers in nearly all energy transformations in the cell. (p. 103-104)
9. The function of enzymes is to catalyze metabolic reactions. The rate of enzyme reactions can be regulated by feedback inhibition, allosteric regulation, competitive inhibition, and noncompetitive inhibition.(p. 105-106)
10. Nutrients cycle between organisms and the environment and are never "used up". Energy flows from the sun to plants through photosynthesis and then to animals as food with large amounts of energy lost during each

conversion. Energy is lost as heat as it flows through the ecosystem. (p. 107–8)

Constructing a Concept Map

Your map should incorporate headings from the chapter and show the relationships between main concepts and subconcepts.

Fill-in-the-Blanks

1. bioenergetics
2. calories
3. kinetic
4. energy
5. chemical
6. metabolism
7. free energy
8. entropy
9. less
10. equilibrium
11. reduced
12. Oxidation
13. anabolic
14. ATP
15. phosphorylation
16. cofactors
17. protons, electrons
18. energy of activation
19. feedback inhibition
20. competitive inhibitors

Putting Your Knowledge to Work

1. C. less than 48
2. B. salads of lettuce and tomatoes
3. A. the antibiotic is a competitive inhibitor
4. C. the sun's energy that reaches the earth's surface
5. B. the second law of thermodynamics

Doing Botany Yourself

Each enzyme reaction has an optimum temperature range. Outside that range the reaction proceeds more slowly. If each repetition was done with the same amount of gelatin and pineapple and each runs for one hour, then the enzyme activities in the pineapple samples can be compared. The greater the amount of gelatin that dissolves the greater the enzyme activity.

Even at 0°C the enzyme still shows activity. This means that freezing the pineapple might not be sufficient to stop enzyme action. It also means that once the pineapple reaches temperatures nearer the optimum range for the enzyme, activity may increase so that there may not be any advantage to freezing the pineapple.

Fresh pineapple would be around room temperature, in the range of 20°–30°C. This is the optimum range for the pineapple enzyme.

Canning requires processing under high temperature, which raises the boiling point of the canned food to above 100°C. The enzyme is inactivated by temperatures this high. Advise the client to use canned pineapple or fresh pineapple that has been boiled for several minutes to ``kill," or denature, the enzyme making it unable to degrade protein.

CHAPTER 6 RESPIRATION

After You Have Read the Chapter

1. In plant cells glucose for cellular respiration can be obtained from what sources?
2. Distinguish between substrate-level phosphorylation and oxidative phosphorylation.
3. Give the locations where glycolysis, the Krebs cycle, and electron transport occur in the cell.
4. What are the products of glycolysis?
5. Why is the Krebs cycle also called the citric acid cycle or the tricarboxylic acid cycle?
6. Relate oxidative phosphorylation to chemosmosis.
7. Compare the roles of NAD+ and cytochromes in cellular respiration.
8. If no ATP is generated by alcoholic fermentation, then why do cells retain this metabolic pathway?
9. Why is oxygen required for aerobic respiration?
10. Describe other types of cellular respiration that occur in plant cells besides aerobic and anaerobic respiration.

Constructing a Concept Map

Using the explanation provided in the introduction to your Study Guide, construct a concept map of the ideas in the chapter.

Fill-in-the-Blanks

1. Sucrose is hydrolyzed into ____________________ and fructose by sucrases in the cytosol, vacuole, and cell walls.(p. 112)
2. Degradation of starch by ____________________ releases glucose phosphates. (p. 113-114)
3. Cellular respiration converts the energy stored in glucose to ________________ that can be used to run cell activities.(p. 112-113)
4. Glucose is converted to ________________ during glycolysis. (p. 115)
5. Before glucose can enter glycolysis it must be ____________________. (p. 115)
6. Carbons from glucose enter the Krebs cycle as ______________. (p. 115)
7. When pyruvate is converted to acetyl CoA ____________________ is released. (p. 115-118)
8. Acetyl CoA is added to ____________________ to initiate the Krebs cycle reactions. (p. 115)
9. Substrate level phosphorylation occurs during ________________ and the Krebs cycle. (p. 114-118)
10. The electron transport system enzymes and proteins are embedded in the ____________________ membrane of the mitochondrion. (p. 119)
11. _____________ are the protein electron carriers of the electron transport system. (p. 122)
12. Electrons removed from NADH by NADH dehydrogenase complex are picked up by ________________ and passed to the cytochrome b-c1 complex. (p. 112)
13. The ______________ unit of a cytochrome contains an iron ion that can pick up or release electrons.
14. Cytochrome oxidase passes electrons and hydrogens to ________________ forming ________________. (p. 123)
15. ________________ produces ATP from energy stored in a hydrogen-ion gradient across the mitochondrial membrane. (pp. 123–124)

16. During electron transport protons are moved from the mitochondrial ________________________ to the space between the inner and outer mitochondrial membranes creating a proton gradient. (p. 123-124)
17. As protons pass through ______________________ a phosphate is added to ADP to form ATP. (p. 124-125)
18. Alcoholic fermentation converts pyruvic acid to _______________ and carbon dioxide.(p. 125-126)
19. The metabolism of sugars via the ______________________________ provides precursors for the synthesis of nucleic acids and lignin.(p. 126)
20. Cyanide-resistant metabolism produces _____________, that raises the temperature of plant tissues well above that of the surrounding habitat. (p. 126)

Putting Your Knowledge to Work

1. A chemical has been added to medium containing isolated mitochondria. The chemical blocks the synthesis of ATP, but not the passage of electrons through the ETS. What is a possible explanation of the affect of this chemical?
   A. It is a noncompetitive inhibitor of cytochrome oxidase.
   B. It is making the inner membrane of the mitochondrion permeable to H+ ions.
   C. It is blocking glycolysis.
   D. It is blocking the synthesis of NADH.

2. A botany student sets up two fermentation tubes containing yeast in a dilute solution of sugar. He puts one on the table top in the lab and the other in the refrigerator. After 2 hours he checks the tubes and finds that the one on the table top has generated 15ml of carbon dioxide, while the one in the refrigerator has generated only 1ml of carbon dioxide. When left on the table for two hours the latter tube also generates 15ml of carbon dioxide. Explain his results.
   A. Light is required for fermentation.
   B. A lack of oxygen in the sealed refrigerator inhibited glycolysis.
   C. The rate of respiration is related to the temperature.
   D. There is no rational explanation of his results.

3. A cell physiologist measured the number of ATP produced by plant cells from a variety of substrates. She found that glucose yielded more ATP than succinate. Which of the following explains this result?
   A. Succinate is incompletely metabolized by plant cells.
   B. Succinate directly inhibits substrate phosphorylation in glycolysis.
   C. Succinate enters respiration at the Krebs cycle by-passing the energy-yielding steps of glycolysis.
   D. Succinate is a competitive inhibitor of ATPases.

4. A home beer brewer wants to increase the conversion of carbohydrates by yeast in the mash to alcohol. What should he do?
   A. Bubble air through the mixture to accelerate respiration in the yeast cells.
   B. Warm the mash in an open vat to about 85°F.
   C. Cool the mash to promote fermentation.
   D. Seal the system and pump out accumulating carbon dioxide.

5. Electrons are removed from the electron transport system when they:
   A. passed to an oxygen and water is formed.
   B. are passed to a molecule of ADP to make a molecule of ATP.
   C. are passed to cytochromes in the outer mitochondrial membrane.
   D. by the NADH dehydrogenase complex.

Doing Botany Yourself

A beer manufacturer wants to stimulate fermentation in her brewing vats. You are asked to design an experiment to evaluate different substrates that can stimulate the yeast to ferment to increase their rate of reproduction and the overall rate of beer production.

ANSWER KEY TO CHAPTER 6 STUDY GUIDE QUESTIONS

After You Have Read the Chapter

1. Glucose for cellular respiration is obtained from the enzymatic degradation of sucrose or starch. (p. 112-113)
2. Substrate phosphorylation transfers a phosphate group from an organic substrate molecule directly to an ADP to produce a molecule of ATP. Oxidative phosphorylation produces ATP by the generation of a proton gradient across the mitochondrial inner membrane. The flow of the protons back through ATP synthetase generates ATP from ADP. (p. 114–115)
3. Glycolysis occurs in the cytosol. The Krebs cycle in the mitochondrial matrix and the mitochondrial membrane. The electron transport occurs in the inner membrane of the mitochondrion. (P. 115)
4. The products of glycolysis are puruvate, NADH, and ATP.(p. 115)
5. The first step in the Krebs cycle produces a citric acid molecule. Citric acid has three carboxyl groups and is thus a tricarboxylic acid. (p. 115-118)
6. Oxidative phosphorylation is the synthesis of ATP using the energy stored in an electrochemical gradient. Chemosmosis refers to oxidative phosphorylation that occurs in the mitochondrion(and chloroplast) when a proton gradient is generated across the inner mitochondrial membrane by electron transport. The stored energy is used to produce ATP. Oxidative phosphorylation and phosphorylation by chemosmosis can be synonyms. (p. 119)
7. NAD+ can pick up hydrogens generated by glycolysis, pyruvate conversion to acetyl CoA, and the Krebs cycle. The hydrogens can be carried to the electron transport system where they can be used to generate ATP. Cytochromes are iron-sulfur proteins associated with the inner mitochondrial membrane. They form the core of the ETS serving as electron acceptors and components of proton pumps that generate the hydrogen ion gradient used to manufacture ATP. (p. 112)
8. Accumulating NADH and pyruvate from glycolysis will inhibit glycolysis(feedback inhibition) and glycolysis will stop. Using NADH and pyruvate to produce alcohol removes the inhibitors so that glycolysis can continue and the cells can produce some ATP. (P. 125-126)
9. Oxygen serves as the terminal electron acceptor for the ETS.It receives the electrons being passed through the ETS allowing to keep functioning. (p. 125)
10. The respiration of 5-Carbon sugars in the pentose phosphate pathway produces precursors for other molecules and NADPH used in the reduction of nitrates to ammonia used in amino acid synthesis. Beta-oxidation converts the fatty acids of lipids to acetyl CoA that can enter the Krebs cycle or in other metabolic pathways. Cyanide-resistant metabolism can dissipate excess energy in fast-respiring cells or can generates large amounts of heat in aroid species as an ecological adaptation. Photorespiration induced in the light when the ratio of oxygen to carbon dioxide is high in some plants reduces crop yields. (P. 126-127)

Constructing a Concept Map

Your map should incorporate headings from the chapter and show the relationships between main concepts and subconcepts.

Fill-in-the-Blanks

1. glucose
2. starch phosphorylase
3. ATP
4. pyruvate
5. phosphorylated
6. acetyl-CoA
7. carbon dioxide
8. oxaloacetic acid
9. glycolysis
10. inner
11. Cytochromes
12. coenzyme Q
13. heme
14. oxygen,water
15. Chemiosmosis
16. matrix
17. ATPases
18. acetyl-CoA
19. pentose phosphate pathway
20. heat

Putting Your Knowledge to Work

1. B. It is making the inner membrane of the mitochondrion permeable to H+ ions.
2. C. The rate of respiration is related to temperature.
3. C. Succinate enters respiration at the Krebs cycle by-passing the energy-yielding steps of glycolysis.
4. D. Seal the system and pump out accumlating carbon dioxide.
5. A. passed to an oxygen and water is formed.

Doing Botany Yourself

Fermentation produces carbon dioxide, which in a liquid is easily visible as bubbles. In a closed tube of liquid the production of carbon dioxide will displace the water solution.

1. Make water solutions of yeast and known concentrations of each substrate to be tested in fermentation U-tubes.

2.Place the tubes and the solutions in a warm place.

3. At ten-minute intervals observe the amount of liquid that is displaced in each tube. This is a measure of the amount of carbon dioxide produced.

4. The substrates that produce the most gas in the shortest time are the best substrates for fermentation.

CHAPTER 7 PHOTOSYNTHESIS

After You Have Read the Chapter

1. What principle did Joseph Priestley demonstrate about the relationship between plants and animals?
2. What is he relationship between the wavelength of a photon and its energy content?
3. What property of pigments is essential in the process of photosynthesis?
4. How does the absorption spectrum for chlorophyll ***a*** relate to the action spectrum for photosynthesis?
5. Where do the reactions of photosynthesis occur in the plant?
6. Describe the roles of antennae complexes, reactions centers, and photosystems in photosynthesis.
7. Compare the photochemical reactions of photosynthesis to the biochemical reactions.
8. How do some plants minimize the effects of photorespiration?
9. What are two environmental factors that control the rate of photosynthesis in a plant?
10. What are some of the possible fates of photosynthates in plants?

Constructing a Concept Map

Using the explanation provided in the introduction to your Study Guide, construct a concept map of the ideas in the chapter.

Fill-in-the-Blanks

1. The "injury" to air observed by early scientist was the removal of _______________ by living animals. (p. 132-133)
2. Jan Ingenhousz showed that _______________ required sunlight to release oxygen. (p. 134)
3. T. W. Engelmann used the alga *Spirogyra* to show that light in the _______________ and _______________ ranges are the best to drive photosynthesis. (p. 135)
4. C. B. van Niel was the first to suggest that the oxygen produced during photosynthesis came from _______________, not carbon dioxide. (p. 135)
5. The energy of red light photon is _____________ than that of a blue light photon. (p. 136)
6. In order for a light photon to excite an electron in chlorophyll it must be _______________ by the electron. (p. 137)
7. Many biological pigments contain a _______________ structure, which holds a metal ion in the center. (p. 137)
8. Photosynthesis is possible in organisms that contain greenish pigments called _______________. (p. 138)
9. _______________ are accessory pigments found in chloroplasts and used by many animals as vitamins. (p. 139)
10. Photosystem II contains P680 and is involved in the production of _______________ from water. (p. 143)
11. Light energy drives ATP synthesis in the chloroplast by _______________ just as they are in the mitochondrion. (p. 143-144)
12. Non-cyclic electron flow during the photochemical reactions of photosynthesis generate _______________, oxygen, and ATP. (p. 146)
13. The final electron acceptor in the chloroplast ETS is _______________. (p. 146-147)
14. The Calvin Cycle is the name given to the series of _______________ reactions of photosynthesis. (p. 148)
15. RuBP carboxylase/oxygenase adds one carbon dioxide to a molecule of _______________. (p. 150)

16. The first products of the Calvin Cycle are two molecules of ___________________________. (p. 150)
17. RuBP carboxylase/oxygenase can release a molecule of _____________________ from ribulose bisphosphate producing a molecule of phosphoglycolic acid and PGA during photorespiration. (p. 150)
18. Some C4 plants can fix carbon dioxide using the enzyme _____________________ to produce oxaloacetic acid from PEP and carbon dioxide. (p. 153-155)
19. Crassulacean acid metabolism allows desert plants to close their stomates during the _________________ and to fix carbon at _____________________ into organic acids. (p. 157)
20. The light intensity at which the rate of photosynthesis equals the rate of respiration is the _____________________ for a plant. (p. 158)

Putting Your Knowledge to Work

1. A plant biochemist used a pH electrode to measure the pH inside the grana sacs of the chloroplasts of a mutant plant during photosynthesis. He found that the pH increased in the grana sacs in the light and decreased in the dark. What is the defect in the plant's chloroplasts?
   A. Photosystem I and Photosystem II are oriented to move protons into the stroma in the light.
   B. The ATP synthetase is moving protons against a proton gradient.
   C. Photosystem I is feeding electrons to Photosystem II.
   D. RuBp carboxylase/oxygenase in the stroma is nonfunctional.

2. A graduate student found that oxygen can be produced by isolated, illuminated chloroplasts using artificial electron acceptors that accept electrons directly from P_{680}. These electron acceptors allow for the release of oxygen without the production of ATP or NADPH. Which of the following would you predict to be true based on this information?
   A. Carbon dioxide is the source of light-generated oxygen by chloroplasts.
   B. Photosystem I can function independently of Photosystem II.
   C. Electrons are moving from Photosystem I to Photosystem II.
   D. Oxygen evolution is associated with Photosystem II.

3. A plant biochemist receives a specimen from a fellow scientist. The scientist noticed that the plant's stomates are closed during the day and under artificial light. He wants the biochemist to analyze the plant to determine how it fixes carbon dioxide under these conditions. The biochemist finds that radioactive carbon dioxide fed to the plants at night is first found in organic acids that accumulate in the vacuole. During the day the label moves to sugars being manufactures in the chloroplasts. If you were analyzing these plants what would you conclude?
   A. The plant is a C4 plant.
   B. The plant is using mitochondria as chloroplasts.
   C. The plant fixes carbon by crassulacean acid metabolism.
   D. The plant is a C3 plant.

4. Red algae grow at depths lower than those to which red and blue light can penetrate in the ocean. What could account for this?
   A. Red algae have accessory pigments that absorb wavelengths of light available at these depths.
   B. Red algae must use infrared energy to power photosynthesis.
   C. Red algae must really be heterotrophs.
   D. The "red alga'" must be misidentified.

5. You are a farmer in Vermont. You are looking for a plant that you can grow efficiently under the 5. apple trees in your orchard to increase your farm production. The orchard is located on well-watered bottomland near a river. The local extension agent suggests a new cultivar of pigweed that is a fast producer due to its C_4 metabolism. You find that the cultivar is overrun by native C_3 pigweeds. What did your extension agent forget?

A. Native species always out-compete foreign species.
B. C3 plants produce toxins that kill C_4 plants.
C. C_3 plants can have the competitive advantage in the shade and in well-watered soil.
D. C_4 plants do not grow well outside the tropics where they evolved.

Doing Botany Yourself

Karen Brueschweiler, an undergraduate at Arizona State University is trying to find the genes for the polypeptides that comprise Photosystem I.

"My part of the project involves tagging the nuclear genes of Photosystem I by inserting a known gene into them. The organisms I use is a mutant of *Chlamydomonas reinhardtii*, a unicellular green alga,. This mutant cannot grow on a culture that lacks the amino acid arginine. When I insert the gene for arginine synthesis, called arg-7, into the mutant, cells that take up the gene can subsequently grow in the absence of supplemental arginine. Devise an experiment to develop mutant strains in which the PS I genes have been interrupted.

ANSWER KEY TO CHAPTER 7 STUDY GUIDE QUESTIONS

After You Have Read the Chapter

1. Although he did not realize it, Priestley demonstrated that plants produce oxygen for animals and that the repsiration of animals generates carbon dioxide for plants. (p. 132-133)
2. The longer the wavelength of a photon, the lower its energy. The shorter the wavelength of a photon, the greater its energy. (p. 136)
3. Pigments are molecules that are able to absorb specific wavelengths of light. (p. 137)
4. The action spectrum of photosynthesis includes those wavelengths of light that support photosynthesis. The absorption spectrum of chlorophyll includes those wavelengths of light that are absorbed by chlorophyll. The aborption spectrum of chlorophyll a most closely matches the action spectrum of photosynthesis,, but not exactly. Other pigments absorb additional wavelengths o light and transfer the energy to photosynthesis. These are called accessory pigments and include chlroophyll b, carotenoids, and xanthophylls. (p. 138)
5. The photochemical reactions of photosynthesis occur in the thylakoid membranes of the grana. The biochemical reactions occur in the stroma or matrix that surrounds the grana. (p. 145)

6. Antennae complexes are aggregates of about 300 molecules of chlorophylla, fifty molecules of carotenoids and other accessory pigments in a protein matrix anchored to the thylakoid membranes. Within the center of the atenna complexes are two chlorophyll a molecules with associated proteins that absorbe energy at a specific wavelength. This is the reaction center. The reaction centers relay electrons directly to one of two photosystems that perform the photochemical reactions of photosynthesis. The P700 reaction center is associated with Photosystem I. the P680 reaction center is associated with Photosystem II. (p. 141-143)
7. Light energy absorbed by Photosystem II chlorophyll causes electrons from chlorophyll to becoome energzed and to be passed to electron acceptors. The electrons are replaced from the dissociation of water and the release of hydrogen ions and oxygen. Light energy absorbed by Photosystem I causes electrons to become energized and passed through a series of electron acceptors to synthesize NADPH. All the photochemical reactions cause the directional transport of hydrogen ions in to the grana sacs. The hydrogen ion gradient is used to drive chemiosmosis and the synthesis of ATP. This ATP and NADPH is used in the Calvin Cycle to synthesize sugars in the biochemical reactions. (p. 145-150)
8. Plants minimize the effects of photorespiration by separating the photochemical and biochemical reactions either by specialized biochemistry, anatomy or by timing of the events. C4 plants have specialized anatomy called Krantz anatomy. The leaf structure acts as a carbon dioxide pump that assures that the carbon dioxide level near chloroplasts carrying out the Calvin cycle is high enough to prevent photorespiration. CAM plants open their stomates at night and fix carbon dioxide into organic acids. In the day the stomates close and the acids release their carbon dioxide. The levels in the leaf are high enough to supress photorespiration. Any carbon dioxide released by photorespiration cannot escape the leaf. (p. 151-152)
9. Environmental factors that control the rate of photosynthesis in a plant are the amount of light available, the concentration of carbon dioxide, the water status of the plant, and metabolic sinks. (p. 158)
10. Photosynthate supports cellular respiration and photorespiration; for amino acid synthesis; stored as starch; to make sucrose for transport; to make cellulose; and to make secondary metabolites. (p. 159-160)

Constructing a Concept Map

Your map should incorporate headings from the text and show the relationships between main concepts and subconcepts.

Fill-in-the-Blanks

1. oxygen
2. plants
3. red, blue
4. water
5. lower
6. absorbed
7. tetrapyrrole
8. chlorophylls
9. Carotenoids
10. oxygen
11. chemiosmosis
12. NADPH
13. P_{700}
14. biochemical
15. ribulose bisphosphate
16. phosphoglyceric acid
17. carbon dioxide
18. PEP carboxylase
19. day, night
20. light-compensation point

Putting Your Knowledge to Work

1. A. Photosystem I and Photosystem II are oriented to move protons into the stroma in the light.
2. D. oxygen evolution is associated with Photosystem II.
3. C. The plant fixes carbon by Crassulacean Acid Metabolism.
4. A. Red algae have accessory pigments that absorb other wavelengths of light available at these depths.
5. C. C_3 plants can have the advantage in the shade and in well-watered soil.

Doing Botany Yourself

To develop strains of algae with interrupted PS I genes you must be able to determine that the arg-7 gene has been inserted into the genes of an algae cell and that the PS I genes are one site of insertion.
1. Expose algal cells to arg-7 genes under conditions that promote uptake of the genes and their insertion into the chromsomes.

2. Plate out these cells onto agar medium lacking arginine. Use a concnetration of algae cells that has been diluted so that each developing colony represents a single cell.

3. Any algal colonies that grow will be those that have the arg-7 gene inserted.

4. Set up an experiment to measure photosynthesis by the colonies using carbon dioxide fixation as a measure of photosynthesis. Those colonies in which the PS I genes have been interrupted will not be able to carry out photosynthesis.

5. These are the colonies you will use to find the location of the PS I genes. Of course, you will needed to supply the cells with a carbohydrate source to keep them alive for any period of time.

CHAPTER 8 PATTERNS OF INHERITANCE

After You Have Read the Chapter

1. What three principles of plant heredity were understood before the time of Mendel's work?
2. State Mendel's theory of heredity?
3. Compare an individual's genotype to its phenotype.
4. Relate Mendel's "factors" to chromosomes.
5. When a cross between a green-seeded pea plant and a yellow-seeded pea plant produces a 3:1 ratio of green seeds to yellow seeds, what can be determined about the character of seed color in peas?
6. What does it mean to be heterozygous for flower color in plants that have either purple or white flowers? (Note: Two white-flowered parents always have white-flowered offspring.)
7. State the chromosomal theory of heredity.
8. How was DNA determined to be the carrier of hereditary information?
9. Describe systems of inheritance that do not involve a dominant and recessive allele.
10. How does linkage and crossing-over affect the inheritance of genetically-determined traits?

Constructing a Concept Map

Using the explanation provided in the introduction to your Study Guide, construct a concept map of the ideas in the chapter.

Fill-in-the-Blanks

1. Gaertner found that ______________________ traits are expressed whenever they are inherited. (p. 167)
2. Mendel used the garden ______________________ in his experiments because it was easy to grow. (p. 167)
3. Each cell of an individual carries two ______________________ for each gene (p. 170).
4. The ______________________ of a pea plant that is homozygous dominant for purple flowers would be purple flowers. (p. 170)
5. When Mendel crossed a pea with white flowers with a pea with purple flowers he was performing a ______________ cross. (p. 171)
6. In a Punnett square the "squares" represent potential ______________________. (p. 171)
7. The F_1 of a cross of a curly-coated guinea pig and a smooth-coated guinea pig gave a ratio of 3 smooth-coated to 1 curly-coated (3:1). The genotypic ratio is__________________. (p. 171)
8. The mating of two peas heterozygous for height and seed color is a ____________________ cross. (p. 171)
9. The parallel behavior of chromosomes and genes led to the development of the ___________theory of heredity. (p. 172)
10. Cells produced by the cell division process of meiosis have half the amount of nuclear genetic material as the parent cells, making them ______________. (p. 172)
11. Experiments with bacteriophages proved that ____________________, not protein, is the genetic material carried in the nucleus of cells. (p. 174-175)
12. A ________________is a sequence of bases that codes for a specific protein or RNA molecule.(p. 175)
13. The ______________________ for the gene determining leaf size in peas would be on the same chromosome in the same position in all normal individuals. (p. 175)
14. Allozymes will show a pattern of inheritance called ______________________ in which both alleles in a heterozygote are expressed equally in an individual. (p. 177)

15. The genes that produce the enzymes in a biochemical pathway are ____________________ genes. (p. 177)
16. Traits controlled by ____________________________ show a range of expression determined by the additive effects of the genes involved and their interaction with the environment. (p. 177-178)
17. Genes that are carried on the same chromosome are said to be ______________.
18. When genes are linked, ________________________________ can produce combinations of traits that do not occur in either parent. (p. 179)
19. Genes carried in the mitochondria and chloroplasts regulate ________________________inheritance. (p. 180)
20. Barbara McClintock discovered_____________________________, DNA sequences that move from one chromosome position to another or to a different chromosome. (p. 180-181)

Putting Your Knowledge to Work

1. A pea plant with smooth, green pods is crossed with itself (self-pollinated). In the F_1 generation all of the pods are green, and 75% are smooth and 25% are hairy. What would be a logical deduction?
   A. Green pods is a dominant character.
   B. Smooth pods is dominant over hairy pods.
   C. Green pods is not a character controlled by genes.
   D. Hairy pods is dominant over smooth pods.

2. A plant breeder took pollen from a pure-breeding, white-flowered d *Arabadopsis* and fertilized the ovules of a pure-breeding orange-flowered plant. All of the offspring were yellow-flowered. The F2 plants produced by self-pollination had a phenotypic ratio of 1(white):2(yellow):1(orange). Flower color in these plants must be the result of:
   A. incomplete dominance.
   B. pleiotropy.
   C. cytoplasmic inheritance.
   D. polygenes.

3. A plant breeder finds the following pattern for albinism in spinach:

Egg		Pollen	Offspring
Albino plant	×	Green plant	All albino plants
Green plant	×	Albino plant	All green plants
Albino plant	×	Albino plant	All albino plants
Green plant	×	Green plant	All green plants

   What kind of inheritance pattern does this suggest?
   A. cytoplasmic inheritance
   B. codominance
   C. pleiotropy
   D. multiple gene inheritance

4. A plant chemotaxonomist is studying a population of sunflowers. She finds that when she does a gel electrophoresis of cellular proteins from a number of individuals in the population that three separate bands appear on the gel showing IAA oxidase activity. When she makes crosses, she finds that plants in the F_1 cannot have more than two of the three forms of the enzyme. The simplest explanation of these results is that the enzymes are
   A. linked.
   B. coded for by incompletely dominant alleles.
   C. allozymes.
   D. isozymes.

5. A plant geneticist crossed a tall, hairy pea plant heterozygous for both characters with a short, smooth pea plant. (Note: Tall and hairy are dominant.) The offspring contained 45% tall, hairy plants; 45% short, smooth plants;10% tall, smooth plants, and 10% short, hairy plants. What did he learn?
   A. The genes are carried on separate chromosomes.
   B. The genes are 20 map units apart on the same chromosome.
   C. The gene is easily mutated.
   D. The genes are part of a multiple allele system.

Doing botany Yourself

A plant geneticist has received seed from a plant breeder for a new type of zinnia. He plants the seeds and gets a mix of yellow- and pink-flowered plants. He wants to determine which color trait is controlled by the dominant allele and to obtain pure-breeding stock. Describe the type of crosses he should make and how to interpret each type of cross.

ANSWER KEY TO CHAPTER 8 STUDY GUIDE QUESTIONS

After You Have Read the Chapter

1. Plant breeders knew that pollen carried inheritable traits, that pollen was required to produce fertile fruits, that some characters present in one generation disappear only to reappear in future generations, and that pure-breeding varieties can be produced. (p. 167)
2. Mendel believed that traits were determined by pairs of factors inherited from each parent. Factors were separated during the formation of gametes and recombined in the formation of new offspring. (p. 168-170)
3. The specific alleles that one carries is one's genotype. One may be homozygous-having two of the same alleles or heterozygous-having two different forms of the allele for a trait. The expression of those alleles is one's phenotype. The phenotype produced by a genotype depends upon the interaction or lack thereof between the two alleles that are inherited and the environment. (p. 170)
4. The behavior of chromosomes during meiosis and fertilization seemed to match that predicted by Mendel's factors. Genes on chromosomes became identified with the factors described by Mendel. (p. 172)
5. Some of the things that are suggested about seed color in peas include: green is dominant and yellow is recessive; the parents are both homozygous for seed color; and one can predict that one out of every four plants produced from this cross will produce yellow seeds and three out of every four will produce green seeds.(p. 171-172)
6. It means that the plant has one allele on one chromosome for purple flowers and one allele for white flowers on the other chromosome of a homologous pair. If *P* is the purple-flower allele and *p* is the white-flower allele, the genotype of this individual would be *Pp*. (p. 170-172)
7. The chromosomal theory of heredity states that chromosomes carry the units of heredity or genes and that the behavior of genes parallels the movements of chromosomes. (p. 172)
8. Bacteriophages with DNA cores and protein coats were used to show that genetic information is transferred to bacteria as DNA while the protein coats remain outside the bacterial cell. (p. 174-175)
9. Other types of inheritance include incomplete dominance, codominance, multiple alleles, multiple genes, epistasis, polygenes, pleiotropy, cytoplasmic inheritance, and transposable elements. (175-179)
10. When alleles for different traits are carried on the same chromosome the traits are said to be linked. Linked genes are normally inherited together. Crossing-over brings together a part of one linkage group on one homologous chromosome with its counterpart on the other homologous chromosome. Chromosome breakage and exchange at these points can lead to new types of offspring and increase the genetic variability of a population. (p. 179)

Constructing a Concept Map

Your map should incorporate headings from the text and show the relationships between main concepts and subconcepts.

Fill-in-the-Blanks

1. dominant
2. pea
3. alleles
4. phenotype
5. monohybrid
6. offspring
7. 1:2:1
8. dihybrid
9. chromosomal
10. haploid
11. DNA
12. gene
13. locus
14. codominance
15. epistatic
16. polygene
17. linked
18. crossing-over
19. cytoplasmic inheritance
20. transposable elements or jumping genes

Putting Your Knowledge to Work

1. B. Smooth pods is dominant over hairy pods
2. A. incomplete dominance
3. A. cytoplasmic inheritance
4. C. allozymes.
5. B. The genes are 20 map units apart on the same chromosome.

Doing Botany Yourself

To determine the dominant and recessive allele the geneticist should first make a number of yellow and yellow, pink and pink, and yellow and pink crosses. From the crosses he will be able to judge which color is dominant.

For example:

1. If yellow × yellow = all yellow or 3 yellow to 1 pink, then yellow is dominant.
2. If pink × pink = all pink or 3 pink to 1 yellow, then pink is dominant.
3. If yellow × pink = all yellow or 50% yellow and 50% pink, then yellow is dominant.
4. If yellow × pink = all pink or 50% pink and 50% yellow then pink is dominant.

Those showing the recessive phenotype will breed true when crossed with themselves or with other homozygous recessives. If the plants are self-pollinating he could allow those expressing the dominant phenotype to self-pollinate. If all of the offspring are of the dominant phenotype then they and their parents can be assumed to be homozygous dominant. Several generations of self-crosses could confirm this.

Otherwise he will have to cross two plants with the dominant phenotype until he gets a cross in which all the offspring show the dominant phenotype. These also will have to go through several generations of crosses to determine that they breed true.

CHAPTER 9 THE CELL CYCLE

After You Have Read the Chapter

1. What is unique about cells in a plant meristem?
2. Divide the cell cycle into three phases?
3. Describe the events that occur in the three phases of interphase.
4. What are the phases of mitosis and how is each identified?
5. Describe the structure of DNA.
6. In what forms does DNA appear in the nucleus?
7. How is new DNA thought to be synthesized from existing DNA?
8. Draw and label a doubled chromosome.
9. Describe cytokinesis in plant cells.
10. What causes the chromosomes to separate during mitosis?

Constructing a Concept Map

Using the explanation given in the introduction to the Study Guide, construct a concept map of the ideas in the chapter.

Fill-in-the-Blanks

1. The __________________ consists of stages of cell growth and cell division. (p. 188)
2. The phase of cell division consists of mitosis and __________________. (p. 188-189)
3. DNA is replicated and chromosomes are duplicated during the ___ phase of interphase. (p. 195)
4. A cell with a chromosome number of four will have eight ______________________ after the S phase of interphase. (p. 203)
5. Cytoskeletal elements, organelle duplication, and cell enlargement occur during ___ phase of interphase. (p. 190)
6. Watson and Crick developed a model for the _________ molecule based on a double helix using the X-ray diffraction analysis data of Rosalind Franklin and Maurice Wilkins. (p. 195)
7. Kinks in DNA molecules in the chromosomes can be removed to allow for replication by enzymes known as ____________________________. (p. 198)
8. In chromosomes the DNA is wrapped around histone complexes to form ____________, the basic unit of chromosome structure in eukaryotes. (p. 193)
9. ____________________ produces two new DNA molecules by adding new nucleotides the old strands of the parent molecule. (p. 195)
10. A new DNA molecule is synthesized ______________________ adds nucleotides to the growing molecule in the 5′ to 3′ direction. (p. 198)
11. The block of DNA between two replication origins is called a ______________. (p. 200)
12. A replication bubble forms between the two ____________________________ formed by the separated DNA strands and DNA polymerases.(p. 2000
13. The________________, formed from microtubules, separates chromosomes attached to it during metaphase of mitosis. (p. 203)

14. Chromatin condenses into visible chromosomes during ________ phase of mitosis. (p. 203)
15. Each doubled chromosome consists of two ____________________ joined at the centromere. (p. 203)
16. The splitting of the centromere during mitosis to create two chromosomes signals the beginning of _______. (p. 204)
17. The spindle fibers disappear and the nuclear envelopes form around the daughter nuclei during _______________. (p. 205)
18. In plant cells the __________________ forms parallel to the spindle to direct the synthesis of the new cell membranes and cell walls during cytokinesis. (p. 206)
19. The _________________ is formed from dictyosome vesicles containing wall materials during telophase. (p. 206)
20. Kinetochore microtubules ____________________ the chromosomes toward the spindle poles, while the spindle microtubules ____________________ the spindle poles apart. (p. 207)

Putting Your Knowledge to Work

1. A cytogeneticist has followed the changes in the average amount of DNA in the nuclei of onion (*Allium cepa*) root-tip cells over a 24-hour period. His results are as follows:

4 hours	8 hours	12 hours	16 hours	20 hours	24 hours
2C	2C	3C	4C	4C	2C

Based on this data determine which of the following would be a true statement?
   A. Anaphase began at 16 hours.
   B. The S phase of the cell cycle occurred between 8 and 16 hours.
   C. The chromosomes were doubled between 0 and 4 hours.
   D. Prophase occurred between 20 and 24 hours.

2. A Chinese postdoctoral fellow at Harvard examined a slide of rice (*Oryza*) root-tip cells from a plant native to her country. She noticed that in some cases a cell division produced a cell with two nuclei instead of two cells with one nucleus each. Which of the following would *not* explain her observation?
   A. The spindle failed to assemble properly.
   B. The centromeres did not split.
   C. The spindle microtubules did not separate the chromosomes.
   D. The cell plate did not form during telophase.

3. A yeast mutant was discovered in which radioactively-labeled thymidine was incorporated into the nucleus, but the cells did not divide. Upon examination it was discovered that over a 24-hour period the size of the nucleus increased until it ruptured, killing the cell. A possible explanation for these observations is that the cells:
   A. are stuck in G1 of interphase.
   B. fail to exit properly from S phase.
   C. lack active DNA polymerases.
   D. do not produce endogenous cholchicine.

4. A mutant petunia is discovered in which the leaf cells have small numbers of chloroplasts and mitochondria. An examination of the cell cycle shows that it is 10% shorter than in non-mutant petunias. A cytologist suggested that the mutation had eliminated a phase of the cell cycle. Which one is it?
   A. telophase
   B. S phase
   C. prophase
   D. G_2 phase

5. A species of unicellular green alga reproduces only by cell division. It is grown in culture labeled with radioactive thymidine. How many generations will it take to find cells in which both strands of the DNA molecules in its chromosomes contain radioactively-labelled thymidine?
   A. four
   B. three
   C. two
   D. one

Doing Botany Yourself

Your thesis advisor wants to determine the average distance of the quiescent center from the tips of roots of corn seedlings. She wants you to devise an experimental method for locating the quiescent center in root tips. Describe what you would do.

ANSWER KEY TO CHAPTER 9 STUDY GUIDE QUESTIONS

After You Have Read the Chapter

1. Cells in a plant mesistem never stop dividing. Meristematic cells complete the cell cycle. (p. 188)
2. The cell cycle consists of interphase, mitosis(nuclear division), and cytokinesis(division of the cytoplasm). Usually mitosis is followed directly by cytokinesis. (p. 189)
3. G1 phase follows cell division and is a period of cell growth and organelle production. The cytoskeleton reforms, organelles multiply, enzymes, proteins, and nucleotides for DNA synthesis are synthesized. Cells may remain in G1 indefinitely until the signal for cell division is received. Then the cell enters S phase during which DNA is replicated, histones are synthesized, and the chromosomes are duplicated. As S phase ends the cell enter the G2 phase during which tubulin for the spindle microtubules, and proteins needed for processing chromosomes and braking down the nuclear envelope are made.
4. Mitosis is a continuous process, but it is divided into phases based on certain observable events. The dissolution of the nucelar membrane, the condensation of the chromosomes, and the formation of the spindle mark prophase. Metaphase begins as the chromosome take position at the metaphse plate of the spindle. Anaphase is marked by the splitting of the centromeres of the chromosomes and their movement toward the spindle poles. Telophase is marked by the chromosomes reaching the poles, the nuclear membranes forming around the daughter nuclei and the chromatin reappearing. If cytoknesis is to follow mitosis the cell plate will be visible during telophase as the phragmoplast and cell plate form the newmembranes and cell walls of the daughter cells.(p. 201-206)
5. DNA is a double helix composed of two anti-parallel strand formed from pairs of four nucleotide bases: adenine/thymine or guanine/cytosine. The nucleotide base sequence in each strand is the genetic code.
6. DNA molecules in the chromosomes are wrapped around cores of histone proteins to form nucleosomes. Stretches of DNA between nucleosomes, spacer DNA, is bound to another histone. These histone bind nucleosomes together to form chromatin fibers that are packed into looped domains that extend from the axis of the chromosome. Some looped domains are coiled further to form chromatin that is visible in the light microscope as heterochromatin. Heterochromatin is thought to not contain functioning genes. Euchromatin, not visible in the light microsocpe, is thought to represent the active DNA.(193-195)
7. New DNA is made by the process of self-replication or semiconservative replication. Helicases untwist the helix and separate the strands. Single-stranded binding proteins prevent the strands from reassociating while DNA polymerases begin the formation of new covalent bonds between nucleotides being added to the old stands using the sequence of nuclotides in the old strand as a template. One strand is synthesisized continually and the anti-parrallel strnad is synthesized in small fragments that are then stuck together. Topioisomerases keep the synthesis moving along by preventing kinks in the moleucle, braking and reforming bonds where necessary. Synthesis continues to the end of the strands and the release of two new DNA molecules
8. Draw and X with a darkened circle where the arms meet. Label the upper and bottom left arms as one chromatid. Label the darkened circle as the centromere. (p. 216)
9. Mitosis is the division of the chromosomes to form two new nuclei. Cytokinesis is the division of the cytoplasm of the parent cell to form two new daughter cells. (p. 201-206)
10. DNA polymerase is responsible for the covalent bonding of the nucleotides in the DNA chain as mandated by the template. (p. 191-192)

Constructing a Concept Map

Your map should incorporate headings from the chapter and show the relationships between main concepts and subconcepts.

Fill-in-the-Blanks

1. cell cycle
2. cytokinesis
3. S
4. Chromatids
5. G_1
6. DNA
7. topoisomerases
8. nucleosomes
9. semiconservative replication
10. DNA polymerase
11. replicon
12. replication forks
13. spindle
14. prophase
15. chromatids
16. anaphase
17. telophase
18. phragmoplast
19. cell plate
20. pull, push

Putting Your Knowledge to Work

1. B. The S phase of the cell cycle occurred between 8 and 16 hour
2. B. The centromeres did not split.
3. C. do not produce endogenous cholchicine.
4. D. G_2 phase
5. C. two

Doing Botany Yourself

During DNA replication radioactive thymidine is incorporated into the cell only in the nucleus. Expose the root tips to the radioactive thymidine for a few hours to allow it to be absorbed by the cell into the nucleus. Then take samples at 1-hour intervals afterwards over a 24-hour period, and make sections through the root-tip meristems for microscopic examination. Mount the sections on slides and coat the slides with a photographic emulsion that is sensitive to radiation given off by the thymidine. The cells in the meristem near the tip of the root that are dividing during that period will take up the radioactive thymidine and incorporate it into the DNA of the nucleus. These cells will show up with black dots in the nuclei. The meristematic cells should be clustered right behind the cells of the root cap.

Those cells in the meristematic region that have not divided will not have dark nuclei. These are the cells of the quiescent center. They should appear as a cluster near the center of the meristem. Using a micrometer attachment on your microscope slide you can calculate the distance to the quiescent center from the tip of the root. Repeating this analysis with many roots will give you the average distance between the quiescent center and the root tip.

CHAPTER 10 MEIOSIS, CHROMOSOMES, AND THE MECHANISM OF HEREDITY

After You Have Read the Chapter

1. Contrast the roles of spores and gametes in plant reproduction?
2. Contrast pollination and fertilization in flowering plants?
3. Describe the products of double fertilization in flowering plants
4. Why does meiosis require two nuclear divisions?
5. Compare Metaphase I and Anaphase I with Metaphase II and Anaphase II.
6. What are some adaptive advantages of polyploids over their diploid parents?
7. Describe the events of synapsis?
8. How does meiosis increase genetic recombination?.
9. Describe three lines of evidence that support the occurrence of crossing-over during meiosis..
10. What are the two most popular theories of the evolution of sexual reproduction?

Constructing a Concept Map

Using the explanation given in the introduction to the Study Guide, construct a concept map of the ideas in the chapter.

Fill-in-the-Blanks

1. Pollen is produced by the meiosis division of ________________________________. (p. 212-214)
2. In flowering plants one to four megaspores form the ________________________ that bears the egg and polar nuclei.(p. 212)
3. A rosebush is an example of the ______________________ generation of the rose plant. (p.212)
4. Division of the generative cell in the pollen produces two ______________________. (p. 214)
5. The _________________________ carries the sperm to the egg and polar nuclei. (p. 214-215)
6. ________________________ chromosomes carry the same genes at the same loci. (p. 215)
7. In Prophase I of meiosis the chromosomes undergo condensation and ______________________. (p. 216)
8. During Meiosis I the _____________ _____________ separate resulting in the production of two haploid nuclei. (p. 216-219)
9. ***Solanum tuberosum***, the Irish potato, is a tetraploid with _______________ sets of chromosomes. (p.219)
10. During Meiosis II the _____________________ of the chromosomes split and the single chromosomes move to the opposite poles. (p. 218)
11. ________________________ are sites of crossing-over and assist in the proper alignment of chromosomes during Metaphase I. (p. 218-220)
12. At the completion of meiosis ___________________________ daughter cells have been formed from the parent cell. (p. 218)
13. Chromosomes are duplicated prior to the _________________________ of meiosis. (p. 216)
14. During synapsis pieces of homologous chromosomes are exchanged by __________________. (p. 219)
15. The exchange of pieces of chromosomes occurs in specific regions of the __________________. (p. 219)

16. The separation and reaasortment of chromosomes during gamete formation leads to a high level of ____________________________ in offspring of sexually reproducing parents. (p. 220-221)
17. In a heterozygote the process of ______________________ can change an allele on one chromosome to the allele on the homologous chromosome. (p. 223-224)
18. Unequal crossing-over can lead to ________________________ expressed as two or more copies of a gene on one chromosome. (p. 224-225)
19. One theory suggests that __________________________ developed as a mechanism to repair damaged DNA. (p. 226-227)
20. The selection for the reproduction of ____________________________ may have led to the development of sexual reproduction. (p. 227)

Putting Your Knowledge to Work

1. A fungal geneticist crossed a grey-spored Neurospora with a green-spored Neurospora. The spore sacs that resulted from the cross contained 4 grey and 4 green spores each. The spores are formed in a single row that reflects the position of the nuclei as they are produced during meiosis and a subsequent mitosis. Assuming that the two alleles are at the same locus and one cross-over has occurred, what is not a possible order for the spores in the sac.
   A. 2 green, 2 grey, 2 green, 2 grey
   B. 4 grey, 4 green
   C. 2 grey, 4 green, 2 grey
   D. 2 green, 4 gray, 2 green

2. A cell has a nucleus with a chromosome number of 2n=16. After meiosis each of the 4 daughter cells will have:
   A. 4 chromosomes.
   B. 16 chromosomes.
   C. 32 chromosomes
   D. 8 chromosomes

3. A plant breeder removes pollen from a plant that is homozygous for purple flower color(P=purple and p= white) and heterozygous for height(T=tall and t=short). what percentage of the pollen can be predicted to have purple and tall alleles?
   A. 25%
   B. 50%
   C. 75%
   D. none

4. A plant breeder finds a petunia that has a red and white striped flower in a test plot. She crosses the plant with a pure-breeding red-flowered petunia. She gets a phenotypic ratio of 1 red: 1 striped. She then crosses plants with striped flowers and gets ration of 1 red: 2 striped: 1 white. She concluded that the allele for red flowers must be an example of :
   A. blended inheritance.
   B. codominance with the white allele.
   C. pleiotropic
   D. incomplete dominance of the red allele over the white allele.

5. A plant geneticist wants to develop a tall pea plant that is resistant to disease using short pea plants that are disease-resistant. The geneticist knows that shortness is caused by a mutated tall gene that produces only half the gene product of the tall gene. What types of genetic recombination events could produce a tall, disease-resistant strain of peas?
   A. gene conversion followed by crossing-over
   B. transposon reassortment
   C. unequal crossing-over and gene duplication
   D. random assortment

Doing Botany Yourself

A mycologist wants to determine if a gene for the production of spiked spores(S=spiked; s=smooth) is linked to a gene for spore color(B=brown; b=white). Devise an experiment that will give her an answer.

ANSWER KEY TO CHAPTER 10 STUDY GUIDE QUESTIONS

After You Have Read the Chapter

1. Spores are produced by meiosis in the sporangium of the sporophyte. Spores develop into haploid plants, gametophytes, that produce gametes by mitosis. Gametes fuse to form a zygote that develops into the sporophyte. (p. 212)
2. Pollination is the transfer of pollen from anther to stigma. Fertilization is the fusion of two gametes. (p. 213)
3. During double fertilization one sperm fuses with the egg to form a zygote that develops into the embryo while a second sperm fuses with the polar nuclei to form the endosperm. The endosperm either provides food for the embryo during development or for the seedling during germination depending on the species. (p. 214-216)
4. The first nuclear division reduces the chromosome number from diploid to haploid. The second nuclear division splits the doubled chromosomes into single chromosomes. The second division assures that in the zygote the nuclear material will be restored t its normal diploid amount. (p. 216-218)
5. During Metaphase I the paired homologous chromosomes line up on the spindle equator. Each member of the pair moves toward an opposite pole during Anaphase I. In Metaphase II the doubled chromosomes line up in a single file on the spindle equator. During Anaphase II the centromeres split and the one single chromosome from each doubled chromosome moves towards the opposite pole. (p. 216-218)
6. Polyploids can form fertile hybrids with other polyploids or other diploids. These "instant species" can express a wider variety of phenotypes allowing the hybrids to adapt to a wider range of environmental conditions than the parents and increasing their survivability. (p. 219)
7. During synapsis protein bridges, synaptonemal complexes, form between homologous chromosomes. These appear as chiasmata under the microscope. (p. 219)
8. Meiosis increases genetic recombination in a variety of ways including: crossing-over that can activate or inactivate genes or convert one allele to another; unequal crossing-over and gene duplication that can create tandem repeats and multiple gene families. Meiosis also makes possible independent assortment of chromosomes during gamete formation and recombination during fertilization that also increase genetic recombination. (p. 220-221)
9. Crossing-over is supported by the observations that movements of identifiable pieces of chromosomes coincide with changes in phenotype; spore analysis of ascomycete fungi; and gene conversion. (p. 222-224)
10. Two theories of the evolution of sexual reproduction are the DNA repair theory and the transposon reproduction theory. The first developed out of the observation that as DNA is replicated mistakes occur that damage the code. Crossing-over allows for the restoration of the code using the undamaged strand. Transposons seem to play a vital role in genetic recombination based on their ability to spread through the genome. Sexual reproduction allows for the reproduction and dispersal of transposons in the genome. (p. 226-227)

Constructing a Concept Map

Your map should incorporate headings from the chapter and show the relationships between main concepts and subconcepts.

Fill-in-the-Blanks

1. microspores
2. embryo sac
3. sporophyte
4. sperm
5. pollen tube
6. homologous
7. synapsis
8. reduction division
9. four
10. centromeres
11. synaptonemal complexes
12. four
13. prophase I
14. Recombination nodules
15. synaptonemal complexes
16. genetic recombination
17. gene conversion
18. gene duplication
19. sexual reproduction
20. transposons

Putting Your Knowledge to Work

1. B. 4 grey and 4 green
2. D. 8 chromosomes
3. C. 75%
4. D. incomplete dominance of the red allele over the white allele
5. D. A or B could give the desired result

Doing Botany Yourself

Neurospora produces a haploid mycelium that cane be grown and crossed with compatible mating strains under controlled conditions. The cross that needs to be made in this case is one between a strain carrying dominant alleles fro each trait and one carrying recessive alleles for each trait.

If the alleles are not linked then the alleles will assort independently producing spores of all four possible phenotypes (brown and spiked, white and spiked, brown and smooth, and white and smooth) in roughly equal numbers.

If alleles are linked, then half the spores will be spiked and brown and half will be white and smooth. If crossing-over has also occurred then a small percentage of non-parental phenotypes(brown and smooth and white and spiked will occur.

CHAPTER 11 MOLECULAR GENETICS AND GENE TECHNOLOGY

After You Have Read the Chapter

1. Explain how instructions for protein synthesis carried in the genes in the nucleus get to the site of protein synthesis in the cytoplasm.
2. Recite the central dogma of molecular biology.
3. Compare transcription to translation.
4. In what ways are RNA transcripts modified to produce functional RNA molecules?
5. What are the steps in the process of transcription?
6. What is the genetic code?
7. Compare the structure of a eukaryotic gene to that of a prokaryotic gene.
8. What types of enzymes are required to engineer a gene?
9. What techniques are used to analyze plant genomes?
10. What kinds of changes in plants are being investigated using genetic engineering techniques?

Constructing a Concept Map

Using the explanation given in the introduction to the Study Guide, construct a concept map of the ideas in the chapter.

Fill-in-the-Blanks

1. Proteins are synthesized on ______________ in the cytoplasm of the cell. (p. 232)
2. The genetic code contained in a gene is transcribed into a molecule of ______________, which is transported to the ribosomes where it directs protein synthesis. (p. 233)
3. New RNA molecules are synthesized by ________________, which initially binds to DNA at a ______________________. (p. 233)
4. Ribosomes are cytoplasmic organelles formed from protein and ________________________ transcribed from nuclear and nucleolar genes. (p. 237-238)
5. The addition of a poly-A tail completes the synthesis of the ____________________. (p. 233)
6. The transcribed region of the gene may consist of a leader sequence, followed by the ______________, the gene, and a trailer sequence. (p. 233)
7. Each ________________________ binds a specific amino acid that is attached at the amino acyl receptor site. (p. 236)
8. The first codon in every plant gene codes for the amino acid ____________________ . (p. 240)
9. Translation terminates when the ribosome reaches a ________________ codon. (p. 240)
10. Each ____________________ in the DNA of a gene consists of three nucleotides.
11. Eukaryotic genes contain noncoding regions called ______________ that interrupt the coding regions called ______________. (p. 242-243)
12. The introns of large primary RNA transcripts are removed by ________________________ formed by complexes of proteins and snRNPs. (p.235)

13. The ______________________ hypothesis suggests that new genes arise by joining independent exons into new combinations creating a protein with a novel function. (p. 243-244)
14. A ___________ transports DNA from the genome of one organism to the genome of another. (p. 245)
15. Cloned DNA fragments from the genome of an individual form a ______________________. (p. 246)
16. ______________________ formed from yeast chromosomes and fragments of foreign DNA make useful tools for studying eukaryotic genes. (p. 247)
17. Specific sequences of DNA (genes) can be located in the genome using radioactive _______________ that can bind to the region of DNA containing the desired genes. (p. 247-248)
18. A technique using ________________________ allows small amounts of a DNA gene-coding sequence to multiply many times in hours instead of weeks. (p. 248)
19. A _______________ plant contains genes from another source in its cells. (p. 250)
20. The ______________________ from *Agrobacterium* can be used to insert genes into dicots but not monocots to create transgenic plants. (p. 250)

Putting Your Knowledge to Work

1. EcoRI has been used to cut a piece of bacterial plasmid DNA to insert a gene into the plasmid. One cut end of a restriction fragment is shown below.

   AATTCATTTCCGCATTAC---------
   GTAAAGGCGTAATG---------

   What must be the nucleotide sequence on the "sticky" end of the gene to be inserted to produce a usable vector?
   A. TTAA
   B. CTTAA
   C. GAATT
   D. AATT

2. A plant molecular biologist has determined that the gene for leaf color and the gene for flower color code for different proteins. The genes that code for the proteins both map to the exact same place on chromosome 4. Which of the following is a possible explanation?
   A. The primary RNA transcript has one set of introns cut out in leaf cells and a different set in flower cells.
   B. The leaf mRNA transcript is copied in the 3′ to 5′ direction while the flower mRNA transcript is copied in the 5′ to 3′ direction from the complementary strand.
   C. The gene spontaneously mutates in all flower cells as they differentiate in bud primordia.
   D. The leaf gene is transcribed by reverse transcriptase.

3. The chloroplast DNA segment containing the gene for P_{680} is cut by three restriction enzymes.
   a. R-1 produces a 10 kilobase (kb) fragment.
   b. An R-1 plus R-2 digest produces a 2 kb, 3 kb, and a 5 kb fragment.
   c. An R-1 plus R-3 digest produces a 4 kb and a 6 kb fragment.

   The 2 kb, 3 kb, and 5 kb are placed into separate wells of an agarose gel and the gel ran. Radioactive probes of the 4 kb fragment bind to the 2 kb fragment and 3 kb fragment positions on the gel. Radioactive probes of the 6 kb fragment bind to the 3 kb and the 5 kb fragment positions on the gel. What is the position of the 2 kb fragment on the 10 kb fragment?
   A. The 2 kb fragment is on the end of the 10 kb fragment.
   B. The 2 kb fragment is to the right of the 3 kb fragment.
   C. The 2 kb fragment is between the 5 kb and 3 kb fragments.
   D. The 2 kb fragment is beside the 5 kb fragment.

4. A new bacterium that lacks DNA polymerase has been discovered. It also has an RNA polymerase that can make a continuous transcript of the bacteria chromosome. What other enzyme would it need to be able to reproduce?
   A. peptidyl transferase
   B. reverse transcriptase
   C. ribosomes
   D. splicozymes

5. A sequencing gel produced from a chromosome segment shows a 3,000-nucleotide base sequence for a gene. A cDNA made from an mRNA transcript produces a 2,000-nucleotide base sequence. Why the difference?
   A. The cDNA sequences are produced from processed RNA transcripts.
   B. The exons in the gene sequence have been spliced out of the cDNA.
   C. The cDNA is made from the primary RNA transcript.
   D. The mRNA lacks a poly-A tail.

Doing Botany Yourself

A plant taxonomist has discovered a new species of goldenrod. He wants to know to which known species it is most closely related. Suggest a way to determine the closest relative of the new goldenrod.

ANSWERS TO CHAPTER 11 STUDY GUIDE QUESTIONS

After You Have Read the Chapter

1. The nucleotide sequences in the genes in the nucleus are used to synthesize mRNA molecules. The mRNA molecules are moved through the pores of the nuclear envelope to the ribosomes in the cytoplasm. The mRNA molecules bind to the ribosomes where they direct the addition of amino acids in a specific sequence to make protein molecules. (p. 235-238)
2. The central dogma of molecular biology is that one gene codes for one protein.Information held in the nucleotide sequences of DNA molecules in the chromosomes is transcribed into RNA molecules that are edited and processed and moved to the cytoplasm as tRNA, mRNA, or as part of the ribosomes. MRNA is translated into polypeptides using tRNA and the ribosomes. These polypeptides are modified, processed, and assembled into functional proteins or enzymes. (p. 235-244)
3. Transcription is the synthesis of RNA(mRNA, tRNA,or rRNA) from DNA and takes place in the nucleus using the gene as a template. Translation is the process of the synthesis of protein using the code carried in mRNA and the assistance of rRNA combined with proteins to construct ribosomes to which the mRNA binds and tRNA molecules that bring the correct amino acids to the site on the ribosomes at which they are added to the growing protein in the correct position. (p. 233-241)
4. RNA transcripts are produced by transcription whose three steps are: initiation, elongation, and termination. (p. 233)
5. The RNA molecule is capped during synthesis with a GTP molecule that protects it from attack by degrading enzymes. A smaller trailer sequence is cleaved from the new RNA molecule to release it from the DNA template. A poly-A tail of 100–200 adenylic acid molecules is added immediately after synthesis. The primary RNA transcript may be cleaved to remove introns. (p. 235-238)
6. The genetic code is a set of codons of three nucleotides each that specify a sequence of specific amino acids in a polypeptide. (p. 241-243)
7. Prokaryotic genes consist only of coding regions, and the RNA transcribed from them requires no processing to form functional RNA transcripts. Eukaryotic genes consist of noncoding regions(introns) that separate the coding regions(exons) of the gene. After RNA is synthesized the portions corresponding to the introns must be removed and the portions corresponding to the exons spliced together RNA processing to produce functional RNA transcripts. (p. 242-243)
8. Enzymes used to engineer genes include restriction enzymes, DNA ligases, and reverse transcriptases. A restriction enzyme is a bacterial enzyme that can cut DNA at a specific nucleotide sequence. If a restriction

enzyme is used to cut bacterial DNA and a gene to be inserted into the bacterial DNA the "sticky ends" will fuse forming a piece of recombinant DNA. The sugars and phosphates of the new chain portion are firmly attached by DNA ligases. When mRNA is available for a desired gene reverse transcriptase is used to make the desired DNA of the gene(cDNA). (p. 244-246)

9. Techniques to analyze plant genomes include: cloning, gene insertion, creation of DNA libraries, creation of artificial chromosomes, radioactive gene probes, PCR, Northern and Southern Blotting, antisense technology, and creatiion of transgenic plants. (p. 245-254)
10. Many possibilities are being investigated including herbicide-resistant crops, disease-resistant crops, insect-resistant crops, drought-resistant crops, changes in storage proteins of food sources, and improvement in the shelf life of fruits. (p. 251-254)

Constructing a Concept Map

Your map should incorporate headings from the chapter and show the relationships between main concepts and subconcepts.

Fill-in-the Blanks

1. ribosomes
2. messenger RNA (mRNA)
3. RNA polymerase, promoter site
4. ribosomal RNA
5. spliceosomes
6. promoter
7. tRNA
8. methionine
8. stop
9. stop or termination
10. codon
11. introns, exons
12. spliceosomes
13. exon shuffling
14. vector
15. genomic library
16. YACs
17. probes
18. polymerase chain reaction (PCR)
19. transgenic
20. T_i plasmid

Doing Botany Yourself

One approach would be to choose a common gene such as rbcL (gene for subunit of RuBP carboxylase/oxygenase). Sequence the gene in all potential close relatives and compare the sequences to that of the new species. The species with the lowest percentage of nucleotide substitutions compared to the new species could be considered to be its closest relative. Another approach would involve digestion of the DNA of the potential relatives and the new species with restriction enzymes. The fragments could be analyzed by Southern Blotting to compare the migration patterns of the fragments. The species whose fragment pattern most closely matches the new species's pattern could be its closest relative.

Chapter 12 Plant Growth and Development

After You Have Read the Chapter

1. What is the major difference between how plants and animals grow?
2. What are two functions of apical meristems?
3. What are three major tissues produced by transitional meristems?
4. Contrast the organization of the apical meristem of a shoot to that of a root.
5. Name two factors that regulate cell enlargement.
6. Describe the plant cell property of totipotency.
7. How can polarity be established in plant cells, tissues, and organs?
8. What role does assymetric cell division play in the development of algae and in the differentiation of plant cells?
9. Identify three signals that control plant growth.
10. What factors prevent all cells in the plant body from responding to a developmental signal?

Constructing a Concept Map

Using the instructions in the introduction of the Study Guide, construct a concept map of the ideas in the chapter.

Fill-in-the-Blanks

1. Photosynthesis, secretion, and storage functions are usually carried out by ________________ cells in plant tissues. (p. 262)
2. The root apical meristem produces the ________________ towards the soil and new root tissue towards the main axis. (p. 268)
3. The apical meristems of stems and roots are responsible for ________________ growth. (p. 263)
4. Primary growth increases the ________________ of plants. (p. 263)
5. The ________________ develops from tissues of the protoderm. (p. 266)
6. Vascular tissue in the youngest portion of the shoot develops from the ________________. (p. 266)
7. Grass can easily regrow after cutting due to the presence of an ________________ meristem at the base of the grass leaf blades. (p. 263)
8. In a plant meristem the ________________ divide occasionally to produce ________________ that divide many times to provide new tissues. (p.263)
9. The ________________ forms the outer surface of the stem apical meristem. (p. 267)
10. The ________________ of the stem apex consists of the central mother cells, pith-meristem, and the peripheral meristem. (p. 267)
11. Few cell divisions ever occur in the ________________ of the root. (p. 268)
12. Cell enlargement requires an increase in ________________ to force apart the cellulose microfibrils that have been loosened by a series of physiological processes. (p. 268)
13. An ________________ division in a leaf epidermal cell determines which cells will develop into guard cells. (p. 274
14. In order for a wound to heal, surrounding tissue must ________________ and then form new tissue to close the wound. (p. 272)

15. Plant development can be controlled by environmental factors such as gravity that establish _________________ within the whole plant, an organ, or a cell. (p. 272)
16. Auxin moving from young leaves down the stem is thought to play a role in the differentiation of ground tissue into _______________ tissue. (p. 275)
17. The asymmetrical distribution of ions such as _________________ can strongly influence plant growth and development. (p. 274)
18. The rate at which a cell divides or enlarges in a plant organ is partly regulated by the __________________ of the cell in the organ. (p. 275-276
19. Plant hormones direct the differentiation of plant cells that have been set by _______________ and biophysical restraints. (p. 277)
20. Roots, shoots, and leaves can be viewed as plant _______________ produced over and over again by the plant's meristems. (p. 278)

Putting Your Knowledge to Work

1. You work for a horticultural firm that wants to produce a rare shrub by using cell culture. What tissue or cell would you choose as source material?
   A. stem cortical parenchyma
   B. tracheids
   C. sieve tube members
   D. guard cells

2. Sections are cut from a willow branch. The sections are planted in pots of soil in a greenhouse with the shoot end of the section exposed and the root end in the soil. Roots sprout from the root end and shoots from the shoot end. Which of the following would you predict about the sections?
   A. the sections lack the property of polarity
   B. the concentration of auxin in the sections is the same from the shoot end to the root end
   C. the root end will produce shoots
   D. dedifferentiation will be the first step in the process of root and shoot formation

3. You are examining a sample of tissue from a section of stem using an electron microscope. You note that the cell-wall cellulose microfibrils are oriented along the horizontal axis along the inner surface of the plasma membrane. In the living stem the cells would be preparing to:
   A. dedifferentiate.
   B. elongate.
   C. divide long the vertical axis of the cell.
   D. expand

4. A section of pea plant stem is placed in a solution of 1 M mannitol. The solution also has an acid pH that should cause the stem to elongate. No elongation is seen. The stem was removed from this solution, rinsed, and placed in acid solution containing no mannitol. It immediately elongates. What is the best explanation for these observations?
   A. Turgor pressure in the stem cells is required for stem elongation.
   B. The acid pH destroyed the natural auxins in the section, preventing it from elongating.
   C. The mannitol poisoned the stem.
   D. Acid pH does not stimulate stem elongation.

5. You work in a plant histology laboratory. You are asked to locate the peripheral meristem in a stem section. How would you recognize these cells?
   A. The cells are cube-shaped and stain faintly.
   B. The cells have large nuclei and stain faintly.
   C. The cells have dense cytoplasm and stain deeply.
   D. The cells lack visible nuclei.

Doing Botany Yourself

You work for a plant physiologist who studies the differentiation of vascular tissue. She wants you to develop a series of experiments to prove that auxin alone is not responsible for the differentiation of both xylem and phloem. She thinks that sugar moving down with the auxin from young leaves is also involved. You have several callus cultures from *Coleus* stems to use as test material. What would you do?

ANSWER KEY TO CHAPTER 12 STUDY GUIDE QUESTIONS

After You Have Read the Chapter

1. Plants never attain a fixed size (they exhibit open, or indeterminate, growth) and have life spans determined by environmental factors. Plants add new tissue at meristems that contain cells that can divide throughout the life of the plant. These new tissues can adapt to changing environmental conditions. Animals have a programmed developmental plan fixed by genetic limits that determine the length of their growth period (closed or determinate growth) and how long each species of animal can live. Addition of new tissue is limited to replacement of lost tissue or repair. (p. 262)
2. Apical meristems establish patterns of development and provide a continual supply of genetically healthy cells. (p. 263-264)
3. There are three transitional meristems: (1) protoderm, which produces epidermis, (2) procambium, which produces vascular tissues, and (3) ground meristem, which produces cortex and pith. (p. 266)
4. The shoot meristem has an outer layer of cells, the tunica, that divide only anticlinally and an inner core of cells, the corpus, whose cells divide in all planes. Organs such as leaf primordia and bud primordia are produced at the edges of the apex on the surface. Within the corpus are areas of central mother cells, pith-rib meristem cells, and peripheral meristem cells, each with its own role in organ development. The root apical meristem consists of a core of cells just behind the root cap. This core contains the quiescent center, actively dividing cells, and the root transitional meristems. The active meristem produces the root cap that covers the meristem and the body of the root. (p. 267)
5. Cell enlargement is controlled by the processes that loosen the cell wall and maintain a negative water potential in the cell vacuole to drive the uptake of water into the cell. (p. 268-269)
6. Totipotency is the property of a single cell to divide and organize itself into a whole plant because each cell in a plant contains all the genetic information to produce a complete individual. (p. 270)
7. Polarity in plant cells, tissues, and organs is established by the interaction of environmental factors such as light, gravity, and temperature and the plant's genome. (p. 272)
8. The division of a cell to produce one large and one small cell(assymetric division) is the first step in the differentiation of leaf epidermal cells into guard cells, root epidermal cells into root hairs, pollen grains into the male gametophyte, and fern rhizoids and reproductive structures. Other developmental signals such as hormones are usually required to complete the developmental process that was begun by the assymetric division.(p. 274)
9. Signals that control plant growth include electrical currents, hormones, and the position of the cell or tissue. (p. 274-278)
10. The cells that surround the potentially differentiating cells and the time in the life cycle when the signal is received can both influence whether or not a cell can respond to a developmental signal and how it will respond. (p. 279)

Constructing a Concept Map

Your map should incorporate headings from the chapter and show the relationships between main concepts and subconcepts.

Fill-in-the-Blanks

1. parenchyma
2. root cap
3. primary
4. length (height)
5. epidermis
6. procambium
7. intercalary
8. initials, derivatives
9. tunica
10. corpus
11. quiescent center
12. turgor pressure
13. assymetric
14. dedifferentiate
15. generative cell
16. vascular
17. Ca^{2+}
18. position
19. genetic
20. modules

Putting Your Knowledge to Work

1. A. stem cortical parenchyma
2. D. dedifferentiation will be the first step in the process of root and shoot formation
3. D. expand.
4. A. Turgor pressure in the stem is required for stem elongation.
5. C. The cells have dense cytoplasm and stain deeply.

Doing Botany Yourself

To determine if auxin or auxin in combination with one or more other factors may be responsible for the differentiation of vascular tissue in Coleus callus, you might set up the following experiments:

1. Make a series of agar solutions containing increasing concentrations of sugar, auxin, and auxin and sugar. Choose physiological ranges for concentrations. Pour the solutions into shallow trays. After the agar has hardened cut into 1-cm blocks.
2. On different callus cultures place the auxin, sugar, auxin-and-sugar blocks, and controls (plain agar blocks). Over a period of weeks track the differentiation of vascular tissue using histological analysis.
3. If no vascular tissue differentiates, then neither the auxin nor the sugar may be involved. If only xylem differentiates when auxin is present, but not when sugar is present, the auxin must be sufficient for xylem differentiation. If phloem differentiates only when sugar and auxin are present, then both must be required for phloem differentiation.

CHAPTER 13 PRIMARY GROWTH: CELLS AND TISSUES

After You Have Read the Chapter

1. What four basic types of tissues make the primary plant body?
2. What functions of the plant body performed by parenchyma?
3. Compare the structure and function of collenchyma and sclerenchyma tissue.
4. Where do fibers form in the plant and how are they used by humans?
5. What roles do guard cells and the cuticle play in minimizing water loss from plant tissues?
6. What functions do trichomes perform in a plant?
7. Compare the structure and function of phloem and xylem.
8. Compare primary xylem to secondary xylem?
9. What types of secretory structures are found in and on plants?
10. What types of substances are secreted by plant tissues?

Constructing a Concept Map

Using the explanation given in the introduction to the Study Guide, construct a concept map of the ideas in the chapter.

Fill-in-the-Blanks

1. ______________________ cells have only primary cell walls and remain able to divide and dedifferentiate. (p. 284)
2. The primary plant body consists mainly of ______________________ that differentiates from the ground meristem. (p. 284)
3. Growing plant parts contain high concentrations of ______________ cells, which have unevenly thickened primary walls and give strength to the plant body. (p. 285-286)
4. The large air spaces in ______________________ promotes gas exchange and provides increased strength to the tissues for maximum support. (p. 285)
5. Seed coats and other tough plant tissues contain high concentrations of ______________________. (p. 285)
6. Cotton fibers are actually long ____________________ that form on the surface of the seed coat. (p. 293)
7. The ____________________ of the plant is covered by a waxy cuticle that protects internal plant tissues from water loss by evaporation. (p. 289)
8. The ____________________ regulate the movement of gases into and out of the plant by regulating the size of the stomates. (p. 291-292)
9. Fungal pathogens can enter a plant through open ______________________.(p. 292)
10. Grass leaves roll up to conserve water when ______________________ cells in the epidermis lose water and shrink.(p. 292)
11. ______________________ are trichomes on roots that increase water absorption and provide a recognition site and point of entry form symbiotic bacteria like *Rhizobium*. (p. 293-294)
12. Water is moved from the roots to the uppermost portions of a tree by the ____________________. (p. 294)

13. ________________________ are long, narrow cells with tapered ends and bordered pits on their walls that allow for the movement of water between cells. (p. 295-296)
14. The transverse walls of ________________________ are dissolved forming a single tube through which water can move rapidly. (p. 296)
15. The end walls of adjacent sieve tube members form the ____________________ that allow solutes to move through the sieve tubes. (p. 297-298)
16. ____________________ cells in the phloem regulate the loading into and unloading of carbohydrates from sieve tubes. (p. 298)
17. Water is secreted by guttation from ________________________ formed by collections of parenchyma at the ends of veins along leaf edges. (p. 299)
18. Plants like *Atriplex* adapt to high salt levels in the soil by secreting excess salt from ____________________ in the leaves of the plants. (p. 299)
19. Ants and other insects are attracted to sugary solutions that are secreted by ________________ on plant organs such as flowers and leaves. (p. 299)
20. Natural rubbers are formed from __________________, which is produced in laticifers of *Hevea brasiliensis* and other plants. (p. 300)

Putting Your Knowledge to Work

1. A taxonomist is sent a new plant. As she examines the plant she decides that it must be an aquatic species. What might she have found that would have led her to this conclusion?
   A. The plant had substantial amounts of epicuticular wax on its root epidermis.
   B. The plant had many stomates on the surfaces of its leaves, but none on the undersides.
   C. The stem ground tissue consisted largely of densely-packed parenchyma.
   D. The plant surfaces were covered with a thick mat of fine hairs.

2. Tracheids lack what characteristic common to all vessel elements?
   A. Tracheids conduct water for one season and then die.
   B. Tracheids are found only in angiosperms.
   C. Tracheids lack openings on their end walls.
   D. Tracheids lack lignified secondary walls.

3. Which of the following is true of the structure of a mature sieve tube member?
   A. The primary cell wall is missing and has been replaced by the secondary cell wall.
   B. It lacks chromosomes.
   C. The plasma membrane has disintegrated.
   D. Its cytoplasm has been replaced by p-protein and callose deposits.

4. A plant breeder wants to develop a new plant that will attract and retain ant defenders. Which of the following characteristics already possessed by the plant could be selected for to increase the likelihood of attracting and maintaining ant defenders?
   A. flypaper leaves and digestive glands
   B. articulated lacticifers containing latex and alkaloids
   C. extrafloral nectaries in the axils of leaves
   D. resin canals

5. A botanist discovered a new plant growing submerged in the waters of a shallow pond. He is trying to decide if the plant is an alga or a small seedless vascular plant. Which of the following would exclude the possibility that it is an alga?
   A. lack of a cuticle
   B. the presence of stomates in the epidermis
   C. sporangia
   D. chloroplasts in its epidermal cells

Doing Botany Yourself

An FDA lab technician examined a sample of tomato catsup from a generic bottle of tomato catsup. He concluded that the tomato catsup had been adulterated by the addition of sawdust. How might he have concluded this?

ANSWER KEY TO CHAPTER 13 STUDY GUIDE QUESTIONS

After You Have Read the Chapter

1. A plant consists largely of meristems, ground, dermal, and vascular tissues. (p. 284)
2. Parenchyma tissue serves as the storage site for starch and other nutrients and as the site for basic metabolic process such as photosynthesis, respiration, and protein synthesis. (p. 284-285)
3. Collenchyma cells have a primary wall, a functional protoplast, and remain capable of stretching and cell division at maturity. They occur in primary tissues and function as strengthening cells. Sclerenchyma tissue has secondary, lignified walls, no protoplast, are not stretchable or capable of division at maturity. They occur in primary or secondary tissues. (p. 285-286)
4. Fibers a elongated sclerenchyma cells that can be found in single strands in various locations in the plant or in bundles associated with the vascular tissue. Xylary fibers are lignified and are considered to be hard fibers. Extraxylary fibers are not lignified and are considered to be soft fibers. Hard fibers are used to make rope, cords, and coarse fabrics. Soft fibers are used to make fine textiles like linen. (p. 288)
5. The cuticle is a waterproof covering formed from a fatty material called cutin. the cuticle can be covered by deposits of epicuticular wax of varying thicknesses that further decreases water loss by evaporation from plant surfaces. Openings in the epidermis, stomata, occur between two guard cells. As the guard cells take up water they swell are push apart forming the stomate or pore. As the guard cells lose water, the edges move back together and the stomate, or pore, disappears. Stomates open and close in a pattern that responds to turgor conditions in the plant and environmental conditions in a way that maximizes carbon dioxide uptake and minimilizes water loss.(p. 289-292)
6. Trichome functions include: absorption of water and nutrients, protection from the plant from predators, production and storage of secondary compounds, secretion of enzymes in carnivorous plants to digest prey and absorption of nutrients released, and recognition of host by pathogen or symbiont. (p. 293-294)
7. Phloem consists of sieve cells or sieve tubes and companion cells, transfer cells, parenchyma, fibers, and secretory structures. Phloem sieve tube elements have an intact plasma membrane and are joined end to end at sieve plates. Phloem is living at maturity and must be metabolically active to move food and other organic nutrients. Xylem consists of conducting elements (tracheids or vessels), parenchyma, and fibers. Lignified walls of tracheids, vessel members, and fibers provide support for the plant body. Tracheids and vessel elements lack an intact plasma membrane at maturity. Metabolic energy is not required to move water and minerals from the soil to all parts of the plant through the conducting elements of the xylem. (p. 297-298)
8. Primary xylem differentiates from procambium in the primary plant body. Protoxylem forms in the elongating plant body. When elongation stops metaxylem forms. Hoops, bands, or helices form in the secondary walls of protoxylem and metaxylem elements provide strength. Secondary xylem differentiates in tissues produced by the vascular cambium. Wood is formed from accumulations of secondary xylem.(p. 295-297)
9. Secretory structures include nectaries, hydathodes, trichomes, salt glands, secretory cells, canals, ducts, cavities, and lacticifers. (p. 299-300)
10. Plants secrete a variety of materials internally and externally including waxes, latex, amino acids, salts, alkaloids, oils, resins, enzymes, terpenes, water, sugar, and organic acids. (p. 299-300)

Constructing a Concept Map

Your map should incorporate headings from the chapter and show the relationships between main concepts and subconcepts.

Fill-in-the-Blanks

1. Parenchyma
2. ground tissue
3. collenchyma
4. aerenchyma
5. sclerids
6. trichomes
7. epidermis
8. guard cells
9. stomates
10. bulliform
11. root hairs
12. xylem
13. tracheids
14. vessel elements
15. sieve areas
16. companion cells
17. hydathodes
18. salt glands
19. nectaries
20. latex

Putting Your Knowledge to Work

1. B. The plant had many stomates on the surfaces of its leaves, but none on the undersides.
2. C. Tracheids lack openings on their end walls.
3. B. It lacks chromosomes.
4. C. extrafloral nectaries in the axils of leaves
5. B. the presence of stomates in the epidermis

Doing Botany Yourself

Tomato fruits, from which catsup is made, are primary tissue. They would contain only primary vascular tissue, which is quite distinct in structure from secondary tissue. The presence of any secondary xylem tissue in the catsup would be suspicious. At the least it would indicate that some vegetative tissue, stems for example, had made their way into the catsup. Large amounts of secondary tissue would mean that the presence of the secondary xylem was the result of a deliberate addition.

All wood from which sawdust is made is largely secondary xylem. Each type of wood has distinctive vessel elements. These can be identified by their length, width, end-wall angle, perforations, and so on. Oak xylem tissue is easily recognizable. Microscopic examination of the catsup would be necessary to identify the vessel elements as primary or secondary tissue and to determine if they come from the tomato or another plant source.

CHAPTER 14 PRIMARY GROWTH: STEMS AND LEAVES

After You Have Read the Chapter

1. Describe four functions of stems.
2. Describe three basic arrangements of vascular tissue in plants.
3. Compare the location of the cortex and the pith in dicot and monocot stems.
4. How can you tell the difference between a solon, a rhizome, and a tendril?
5. Compare the fleshy portions of bulbs, corns, and tubers.
6. Describe the development of a leaf from a primordium to a mature leaf.
7. Define phyllotaxis and the types displayed by plants.
8. How can you determine if a leaf is simple or compound?
9. What are two environmental factors that can determine leaf form?
10. How have leaves been adapted to different environments and to defend against predators?
11. Describe the types of economically important products that are obtained from leaves.

Constructing a Concept Map

Using the instructions in the introduction of the Study Guide, construct a concept map of the ideas in the chapter.

Fill-in-the-Blanks

1. A consists of nodes where leaves are attached separated by nternodes. (p. 308)
2. In a vascular bundle _____________________ always differentiates on the inside of the bundle. (p. 310)
3. In non-woody plants vascular tissue is organized into _____________________ that form cylinders of conducting tissue from the leaves to the roots. (p. 310)
4. Monocot stems lack a central _________________ region so that their ground tissue parenchyma spread throughout the stem. (p. 308)
5. New shoot growth can be produced from ___________________ formed in the axils of leaves. (p. 311)
6. Cacti have developed succulent stems to store ________________ as an adaptation to desert habitats. (p. 311-312)
7. A marginal meristem forms the _______________ and the petiole of the leaf. (p. 313)
8. A plant with one leaf per node has ___________ phyllotaxis. (p. 314)
9. Phyllotaxis follows a mathematically predictable pattern, a _____________________, that bears the name of its discoverer, an Italian mathematician. (p. 316)
10. Vertical files of leaves on a stem are called _________________. (p. 314)
11. Compound leaves with all the leaflets meeting at one point, such as those of the horse chestnut (*Aeschylus hypocastanum*), are ___________ compound. (p. 317)
12. Leaf epidermal cells lack _________________ that are abundant in the palisade mesophyll. (p. 314)
13. Gas exchange and photosynthesis occur in the _______________________ of the leaf. (p. 318-319)
14. The _____________________ of leaves are formed from vascular bundles. (318-319)
15. _____________________ often have modified leaves with thick cuticles, sunken stomates, and large amounts of sclerenchyma. (p. 322)

16. In response to differences in available light leaves on the same plant move to form _______________ to minimize the shading of leaves. (p. 324)
17. Carnivorous plants use pitcher-shaped leaves and spring-loaded traps to catch _______________, which are sources of nutrients. (p. 326)
18. Some plant leaves contain chemicals similar to insect _______________ that disrupt the normal development. (p. 327-328)
19. Leaves separate from the stem at the _______________ when the cells of the separation layer become suberized. (p. 329)
20. Digitalis from foxglove and caffeine from coffee are plant chemicals that can stimulate the _______________. (p. 330)

Putting Your Knowledge to Work

1. A botanist has found a new plant that has "thorns". True thorns are derived from leaves. How can she tell if the thorns are derived from leaves?
   A. Each thorn will have a bud in its axil.
   B. The thorns will occur only along internodes.
   C. The thorns will consist of a prophyll and a petiole.
   D. The immature thorn will be covered by bud scales.

2. A horticulturalist wants to develop a mail order business. He wants to ship plant material in a form that will produce new plants, but that does not require water and light to survive the trip. which of the following plant organs would not be a good choice?
   A. corms
   B. tubers
   C. rhizomes
   D. cladodes

3. An ecologist samples a population of a species of desert shrub. He finds that the DNA of all the shrubs produce identical DNA fingerprints after gel electrophoresis. What is an explanation for all of the members of a plant population having the same genetic composition?
   A. They all developed from seeds produced in the same year on one parent shrub.
   B. They developed from a seed containing multiple embryos.
   C. They are haploids.
   D. They all developed on rhizomes produced from a single parent shrub.

4. A taxonomist receives a specimen of a new plant. She can tell that this plant lives in water with its leaves floating on the surface. How does she know?
   A. The leaves have stomates only on the lower surfaces.
   B. The leaves have stomates only on the upper surfaces.
   C. The stem epidermis is sclerified.
   D. The plant has an extensive, deep root system to protect against being uprooted by wave action.

5. A plant anatomist examines isolated leaves found at a crime scene. Which of the following features would identify the plant leaf as that of a tropical grass?
   A. uniform chlorenchyma concentrically organized around a photosynthetic bundle
   B. more stomates on the lower surface than on the upper surface
   C. compound leaves
   D. highly differentiated palisade and spongy mesophyll sheath.

Doing Botany Yourself

A student notices that the buds of the lilac tree outside her bedroom window break dormancy and begin to produce new shoots every spring as the weather gets warmer. She also notes that the days are getting longer as well. She wants to determine if days of increasing length or increasing temperature is the basic control for breaking of dormancy in buds. Devise an experiment by which she can do this.

ANSWER KEY TO CHAPTER 14 STUDY GUIDE QUESTIONS

After You Have Read the Chapter

1. Stems support leaves, produce carbohydrates, store materials, and transport water and solutes between roots and leaves. (p. 308)
2. Vascular bundles are either scattered throughout the stem (as in monocots), arranged in a ring around a pith (as in many dicots), or are arranged as continuous rings of xylem each surrounded by a ring of phloem (as in nonflowering plants and some dicots). (p. 310)
3. The cortex is a region of ground tissue parenchyma that lies between the vascular tissue and the outer surface of the stem. The pith is a region of ground tissue parenchyma in the center of the stem often specialized for storage. These regions are present in dicot stems but are less evident or absent in monocot stems because vascular bundles are present throughout the stem. (p. 309-310)
4. Stolons are stems that run along the surface of the soil and are usually specialized for vegetative reproduction as in strawberries and grasses. Rhizomes are stems that run underground near the surface of the soil producing adventitious roots on the underside of the organ. Rhizomes are specialized for food storage and for regrowth of surface shoots after a period of stress like the cold temperatures of winter. Tendrils are shoots with adhesive pads that stick to objects and can coil around objects to provide support. (p. 311)
5. The fleshy portion of a bulb is comprised of leaves modified for storage. The fleshy region of a corm is a stubby, vertically oriented, enlarged stem that develops underground. The fleshy portion of a tuber is a portion of a stem that enlarges underground. (p. 312, 326)
6. The leaf primordium develops as apical peg with an apical meristem and a procambium that will form the midrib. an adaxial meristem thickens the leaf forming an upper leaf zone with a marginal meristem that forms the blade and petiole and a lower leaf zone that forms the leaf base. The marginal meristem adds to the blade and differentiation of the tissues forms a mature leaf. (p. 313)
7. Phyllotaxis refers to the arrangement of leaves on the stem. Plants may have alternate, opposite, or whorled phyllotaxis. A plant that has one leaf per node(point on stem where leaves are attached) has alternate phyllotaxis. A plant with two leaves per node has opposite phyllotaxis. A plant with three or more leaves per node has whorled phyllotaxis. (p. 314-315)
8. A simple leaf consists of a single blade and a petiole with a bud in the axil formed where the petiole meets the stem. A compound leaf consists of a blade divided into many smaller blades, the ``petioles" of which do not have buds in the axils. The bud is where the petiole of the compound leaf meets the stem. (p. 316-318)
9. Two environmental factors that can modify leaf development are light (presence or absence, day length, and intensity) and moisture. (p. 321-324)
10. Xerophytic plants have a number of modifications including small, thick, spongy leaves, an epidermis with thick cell walls, many stomata (often sunken or in crypts), a thick cuticle, and large amounts of supportive tissue in leaves. Hydrophytes have large, thin leaves with aerenchyma and photosynthetic epidermis, a thin cuticle and epidermis, and stomata absent or on upper surface of leaves, and they have little xylem or

supporting tissue in their leaves. Hydrophytes display leaf dimorphism with underwater leaves being highly dissected and air leaves having a flat blade. Other specializations include window leaves of desert plants that allow the plants to photosynthesize underground. Flower pot leaves that collect nutrient-rich debris and insect-trapping leaves allow plants to grow under nutrient poor conditions. Plant leaves produce chemicals that poison animals and insects, hormones that disrupt insect and vertebrate reproduction and development, and antimicrobial phytoallexins. (p. 322-323)

11. Plant leaves are used as sources of food, spices, drinks, dyes, fibers, fuel, drugs, soaps, and creams. (p. 329-331)

Constructing a Concept Map

Your map should incorporate headings from the chapter and show the relationships between main concepts and subconcepts.

Fill-in-the-Blanks

1. stem
2. xylem
3. vascular bundles
4. pith
5. buds(axillary buds)
6. water
7. blade
8. spiral or alternate
9. Fibonacci series
10. orthostichies
11. palmately
12. chloroplasts
13. spongy mesophyll
14. veins
15. Xerophytes
16. mosaics
17. insects
18. hormones
19. abscision zone
20. heart

Putting Your Knowledge to Work

1. C. Each thorn will have a bud in its axil.
2. C. two.
3. B. The plants were responding to the less stressful greenhouse conditions.
4. B. The leaves have stomates only on the upper surfaces.
5. D. uniform chlorenchyma concentrically organized around a photosynthetic bundle sheath

Doing Botany Yourself

During the summer the student could make 30 cuttings from the lilac bush and root them in pots of soil. She could take 10 of the pots inside and place them near her window facing the parent plants. Another 10 could be placed near the parent plants outside, and the third 10 could be put in a large cardboard box covered with foil to prevent light from getting to the plants. These covered plants could be placed in an unheated garage and watered at night to prevent exposure to light.

The plants outside would be exposed to light and freezing temperatures. Those in the room would be exposed to light but not to freezing temperatures. The student must be careful not to turn on her bedroom lights during the winter after sundown so that the plants inside are not exposed to more light than those outside. Those in the garage would be exposed to freezing temperature but no light. Once the plants outside have broken dormancy, the plants in the garage can be placed outside in the light.

If freezing temperature is required for breaking dormancy, then the plants in the room should remain dormant, despite being exposed to days that are increasing in length as spring approaches. If long days are the determining factor, then the plants in the garage will not break dormancy once placed outside despite previous exposure to freezing temperatures.

CHAPTER 15 PRIMARY GROWTH: ROOTS

After You Have Read the Chapter

1. Identify and describe the kinds of root systems found in plants.
2. Describe the functions of roots.
3. Compare the root cap to the subapical region of the root?
4. Contrast the role of the endodermis in water and mineral uptake by roots.
5. Describe the organization of tissues in a mature dicot root.
6. Describe the rhizosphere of a root.
7. How do plant roots deal with competitors for water and nutrients?
8. What factors can alter the pattern of root growth?
9. Describe the adaptive modifications of roots.
10. What types of economically important products can be obtained from roots?

Constructing a Concept Map

Using the instructions in the introduction of the Study Guide, construct a concept map of the ideas in the chapter.

Fill-in-the-Blanks

1. A large ____________________ allows the desert shrub mesquite to reach deep underground stores of water in its desert habitat. (p. 336)
2. Grasses produce a ________________________ root system from adventitious roots that functions to hold soil in place and prevent erosion. (p. 336)
3. Roots absorb large amounts of _________________ and minerals from the soil. (p. 336)
4. The ________________ protects the growing root tip, and senses light, gravity, and pressure to alter root growth. (p. 338)
5. Dictyosomes secrete ______________________, a polysaccharide, to provide protection and lubrication as the root pushes through the soil and to increase water and mineral absorption. (p. 339)
6. The ____________________ in the apical meristem contains cells arrested in G_1 phase that serve as a reserve meristem and sets the pattern of growth in the root. (p. 339)
7. The innermost layer of the cortex of the root is the _________________, which regulates the movement of water and nutrients through the symplast. (p. 341)
8. The _____________________ of a monocot root contains the pericycle, vascular tissue, and a pith. (p. 343)
9. All lateral roots develop from the __________________________ that lies just inside the endodermis and outside the vascular tissue of the root. (p. 343)
10. In roots the xylem alternates with the ______________________________ in the stele. (p. 345)
11. Roots that grow downward into the soil are positively ______________________. (p. 348)
12. Water and dissolved minerals move from the soil into the _______________________ of the plant and travel to the endodermis where they enter the symplast. (p. 343)
13. Light inhibits the rate of root ___________________________by slowing cell division and elongation. (p. 348)
14. Some plants can produce ______________________________ of genetically identical plants covering many acres through adventitious buds called root suckers. (p. 350)

15. Black mangrove (*Avicennia germinani*) provides aeration to root tissues through ___________________, which prevent anaerobic respiration in water-covered roots. (p. 350)
16. The ________________________ roots of lilies and gladiolias pull the bulbs deep into the soil. (p. 350)
17. Roots and fungi can form mutualistic associations called ________________ that increase nutrient absorption and stress tolerance in the plant host. (p. 350-351)
18. *Rhizobium* bacteria infect host roots, resulting in the formation of _______________ on the root in which the bacteria fix nitrogen. (p. 351)
19. The velamen of the __________________ roots of orchids is modified for mechanical protection and prevention of water loss by evaporation from the root surfaces. (p. 351)
20. Supportive aerial roots called ________________________ roots form support systems for shallow-rooted plants such as corn and banyan trees. (p. 352)

Putting Your Knowledge to Work

1. You are given the tip of a plant organ and asked to examine the organ both intact and sectioned under the microscope. Which of the following would indicate that the organ is a root tip?
   A. lateral appendages produced superficially
   B. an apical meristem with a tunica corpus organization
   C. a subapical meristematic region
   D. a well-developed endodermis just under the epidermis

2. Which of the following is not an adaptation that increases nutrient availability to plants through their roots?
   A. aerial roots
   B. mycorrhizal roots
   C. parasitic roots
   D. root nodules containing symbiotic bacteria

3. A plant breeder wants to increase the potential attractiveness of the roots of wheat plants to soil microbes that might develop beneficial relationships with the plant roots. Which of the following would the breeder want to do in developing the plant?
   A. increase mucigel secretion from plant roots
   B. increase phytoallexin secretion by plant roots
   C. decrease the formation of root hairs on the root epidermis
   D. increase retention of carbohydrates in storage parenchyma to prevent leakage to the soil

4. A greenhouse assistant wants to stimulate roots on cuttings he is propagating. Which of the following should give him the best results?
   A. maintain a 20-50% moisture content in the rooting mixture
   B. use a soil mixture that is 100% clay
   C. inject ethylene into the soil once a day
   D. keep the temperature below 50 degrees F to simulate springtime temperatures

5. A farmer tests her soil and finds the pH=5.0 Which of the following can she expect?
   A. Her crops will develop aluminum toxicity.
   B. Root growth will stop.
   C. She will have to add lime to the soil to raise the pH to 7.0 or the plants will die.
   D. As roots take up cations the pH of the soil will decrease.

Doing Botany Yourself

Devise an experiment to test if the relationship between mycorrhizal fungi and the host plant is mutually beneficial.

ANSWERS TO CHAPTER 15 STUDY GUIDE QUESTIONS

After You Have Read the Chapter

1. Taproots are enlarged primary roots specialized for anchorage and storage. Fibrous root systems are widespread, shallow systems developed from many finely branched adventitious roots specialized for quick absorption and holding soil in place. Adventitious roots can develop on rhizomes, serve as reproductive roots, form clones, and develop on cuttings for the commercial propagation of plants. (p. 336)
2. Functions of roots include anchorage to the substrate, storage of food, absorption of mineral nutrients and water, and conduction of water and minerals to other plant regions. (p. 338)
3. The root cap is formed by the apical meristem and protects it. The root cap secretes mucigel that lubricates the root for easier passage through the soil and increases absorption of water and nutrients the
   The subapical region of the root includes the zones of cell division (apical meristem), elongation, and maturation (where elongation ends and cell differentiation begins). (p. 338-340)
4. The endodermis is the innermost layer of the cortex and surrounds the stele. The Casperian strip, a suberized strip in the radial and transverse walls of the cells prevent water and mineral transport through the cell wall. Water and minerals must cross the plasma membrane of the endodermal cells that can regulate uptake. The endodermis regulates movement of water and minerals into the vascular system and prevents leakage from the vascular tissue. (p. 341-342)
5. A dicot root has endogenous lateral appendages, exarch xylem, no pith, and vascular tissue organized into a protostele. The protostele has a solid core of xylem and arms of xylem alternating with patches of phloem with a vascular cambium between the phloem and xylem. (pp. 345)
6. The rhizosphere is the narrow zone of soil surrounding the root, which is modified by growth and metabolic activities of the root. These modifications include enrichment with organic matter, compression of soil particles, and changes in soil pH, carbon dioxide, oxygen, and nutrient content. (p. 346)
7. To minimize competition between root systems plants produce more roots faster than their neighbors, they produce roots in different regions of the soil than their neighbors, and they secrete compounds that inhibit the growth of neighboring root systems. (p. 347-348)
8. Factors that control growth and distribution of roots include temperature, the influences of other organisms, light, gravity, genetics, stage of development, soil texture, soil moisture and aeration, soil nutrients, soil pH, and soil ethylene content. (p. 347-349)
9. Roots can undergo modification to increase storage capacity, to propagate (via organs such as suckers), to provide aeration in water-logged environments (by forming pneumatophores), to pull plant corms or bulbs deep into the soil (using contractile roots), to form connections between a parasitic plant and its host (as in mycorrhizae or fungal-root associations), to form root nodules containing nitrogen-fixing microbes, and to form aerial roots on epiphytes specialized for water retention, photosynthesis, or support. (p. 349–352)
10. Products obtained from roots are foods rich in carbohydrates, drugs, and dyes. (p. 352)

Constructing a Concept Map

Your map should incorporate headings from the chapter and show the relationships between main concepts and subconcepts.

Fill-in-the-Blanks

1. taproot
2. fibrous root
3. water
4. root cap
5. mucigel
6. quiescent center
7. endodermis
8. stele
9. pericycle
10. phloem
11. geotropic
12. apoplast
13. growth
14. clones
15. pneumatophores
16. contractile
17. mycorrhizae
18. nodules
19. velamen
20. prop

Putting Your Knowledge to Work

1. C. a subapical meristematic region
2. A. aerial roots
3. A. increase mucigel secretion from plant roots.
4. A. Maintain a 20–50% moisture content in the rooting mix.
5. D. As roots take up cations the pH of the soil will decrease.

Doing Botany Yourself

The most logical mutualistic relationship between mycorrhizal fungi and the host plant would be for the fungus to supply minerals absorbed from the soil to the host and for the host to supply carbohydrate to the fungus.

You will need host plants not infected with mycorrhizal fungi and host plants with mycorrhizal roots.

Radioactive carbon dioxide fed to a host plant leaf sealed in a chamber will produce radioactively labelled sugars. If these sugars can be detected in the fungus in the roots of the plant, then the fungus is getting carbohydrate from the host.

Radioactively labelled nitrates, phosphates, potassium and other nutrients can be supplied to the roots of host plants infected with mycorrhizal fungi and those not infected. The rates of mineral uptake can be compared. If uptake is significantly higher in the infected roots, the fungi may be assisting in the uptake.

Autoradiographs of infected roots that show concentrations of radioactivity in the zones containing fungi would also suggest the involvement of the fungi in nutrient uptake.

CHAPTER 16 SECONDARY GROWTH

After You Have Read the Chapter

1. What is the location and the function of the vascular cambium in the plant body?
2. Compare a ray initial to a fusiform initial?
3. What factors are involved in controlling the activities of the vascular cambium?
4. How can you distinguish hardwood from softwood?
6. What characteristics of lumber are used to evaluate its suitability for various uses?
7. What tissues are included in the bark of a tree?
8. Where is the periderm located and how does it form?
9. Describe some unusual forms of secondary growth.
10. What are some of the commercial uses for wood?

Constructing a Concept Map

Using the instructions in the introduction to the Study Guide, construct a concept map of the ideas in the chapter.

Fill-in-the-Blanks

1. The ____________________ plant body is produced by the activity of lateral meristems like the vascular cambium and phellogen. (p. 358)
2. The rays in wood are produced by ____________________ initials in the vascular cambium. (p. 358)
3. Vascular cambium produces 4 to 10 times as much ____________________ as ____________________ tissue. (p. 358)
4. The drier days of summer slow vascular cambium activity, producing thicker-walled, smaller cells typical of ____________________ wood. (p. 360-362)
5. Dendrochronologists can determine the age of a tree and the climate history of its habitat from an examination of its ____________________. (p. 362)
6. The ________________ of trees is formed as secondary xylem accumulates over a period of years. (p. 362)
7. Softwoods contain only ________________while hardwoods contain ______________ and vessels. (p. 363)
8. Ring-porous woods have large-diameter____________________ concentrated along growth rings. (p. 364)
9. The ________________ consists of the outer few centimeters of secondary xylem and functions in the transport of water and nutrients between the roots and the leaves. (p. 367)
10. In a ____________________ section of an oak trunk vessels appear as large circles or pores. (p. 366-367)
11. A ______________ forms in the wood when the base of a dead lateral branch becomes covered over by the lateral growth of the main stem. (p. 367)
12. Wood with a ____________________ of 0.5 would be half as dense as water. (p. 367)
13. When woody stems are bent by wind or other forces into a horizontal position, the formation of ____________________ allows the stem to grow vertically again. (p. 369)
14. The____________________ of a tree consists of the secondary phloem and the periderm. (p. 370)

15. The secondary phloem consists of sieve elements alternating with bands of thick-walled ____________. (p. 370)
16. _______________ allow air to reach the internal tissues of the woody stem. (p. 373)
17. The vascular cambium in carrot and cacti produce large amounts of ____________ tissue, which is used as storage tissue. (p. 374)
18. __________________, or pulverized wood, is used to make paper, cardboard, nitrocellulose explosives, and many other products. (p. 376)
19. ___________________________ is made by gluing together many layers of veneer at right angles. (p. 376)
20. Burning hardwood in the presence of little air forms ______________________. Subsequent distillation of the product produces methanol, acetic acid, and rosin. (p. 377)

Putting Your Knowledge to Work

1. A dendrochronologist took a core sample from a tree in a desert in the southwestern United States. The sample lacked discernable growth rings. She was puzzled. What is a possible explanation for the lack of rings?
   A. Growth in the desert occurs at a steady rate throughout the year so growth rings do not form in desert trees.
   B. The ``tree" is an arborescent monocot.
   C. Continual water stress decreases the pore-size difference between spring and summer wood.
   D. The plant lacks a functional phellogen.

2. A plant anatomist is examining a sample of wood taken from a crime scene. He finds that the sample has swirls on its surface, long vertical ``tubes," and stacks of ``circles" dispersed through the sample. He concludes that the sample is a piece of oak from a desktop since flat furniture lumber is usually taken as tangential cuts from a log. How does he know the sample is a tangential cut?
   A. Rays in a tangential cut appear as stacks of circles.
   B. Rays appear as vertical tubes in a tangential cut.
   C. Vessels appear as circles in a tangential cut of wood.
   D. Rays appear as swirls in a tangential cut of wood.

3. A plant anatomist is sent a sample of wood from the FBI. He immediately knows that the sample is probably from a pine or spruce tree. How?
   A. The wood has extensive areas of fibers.
   B. The wood contains only tracheids and vessels.
   C. The tracheid walls are very low in lignin content.
   D. The wood has numerous resin canals lined with parenchyma cells.

4. An orchard owner notices that rabbits eat the bark off young trees in a complete circle around the trunk at snow level. The trees remain healthy through the following growing season, but do not survive the winter. How can you explain his observations?
   A. During the winter ice forms in the vessels of t he xylem "exposed" by the removal of the bark so that the shoot of the tree can no longer get water and the tree dies.
   B. Food stored in the roots over the summer cannot reach the developing buds the next spring and so they fail to develop and the tree dies.
   C. Water loss from the exposed surface exceeds the amount of water absorbed by the roots.
   D. Removal of the bark prevents movement of sugars from the leaves to the roots. Once stored sugars in the roots are used up they and the tree dies.

5. A little boy standing beside a tree in his yard once carved a heart with his and his girlfriend's initials into the bark. Twenty-five years later he returns to the tree. He wants to show the heart to the girl, now his wife. Can he?
   A. No. The heart will be many feet above the ground by now.
   B. Yes. The heart will only be a couple of feet higher than when he carved it.
   C. Yes. The heart will be at the same height as when he carved it.
   D. No. All signs of the heart would have been obliterated by periderm expansion.

Doing Botany Yourself

Vascular tissue differentiates in young stems in parenchyma tissue. Stimuli from the apex and axillary buds seem to be able to control where the vascular tissue forms. How could you set up a series of experiments to determine if substances from developing buds can stimulate the formation of vascular tissue in masses of undifferentiated parenchyma callus cultured from pith tissue.

ANSWERS TO CHAPTER 16 STUDY GUIDE QUESTIONS

After You Have Read the Chapter

1. The vascular cambium is located between the xylem and the phloem in stems. In the young plant body this is within the vascular bundles (fascicular) and between the vascular bundles (interfascicular). In the older plant body a ring of vascular cambium forms between the secondary phloem near the cortex and the secondary xylem toward the pith. (p. 358)
2. Ray initials are small elongate cells in the vascular cambium oriented perpendicular to the stem axis that form the ray parenchyma of the rays that form spokes in the secondary vascular tissue. Fusiform initials are tapered, prism-like cells in the vascular cambium oriented parallel to the stem axis. Fusiform initials produce secondary xylem toward the inside of the stem and phloem toward the outside. (p. 358)
3. The activity of the vascular cambium is regulated by environmental factors such as day length and temperature, which may control hormonal concentrations (auxin and cytokinins) and by genetic factors that determine developmental state. (p. 360–362)
4. Hardwoods form in angiosperm trees in which large concentrations of fibers in the secondary xylem make the wood denser or ``harder." Softwoods form in nonflowering seed plants in which there are no fibers and the woods are less dense and lighter. (p. 363-364)
5. Characteristics of wood that can give clues to its suitability for commercial purposes include cut(cross section, radial section, or tangential section), the presence of knots, grain, texture, density, durability, and water content. (p. 367)
6. The bark includes all the tissues outside the vascular cambium specifically the secondary phloem and the periderm (phellogen, phellem, and phellogen). (p. 370)
8. The periderm is the outer covering of the stem that replaces the epidermis as secondary growth increases the diameter of the stem. It forms from the cork cambium (phellogen) initiated in the inner layers of the cortex. It

produces living cortical cells toward the phloem (phelloderm) and cork tissue (phellem) toward the outer surface. Cork cells have heavily suberized cell walls and lack a living protoplast at maturity. (p. 370, 372-373)

9. Unusual forms of secondary growth include reaction wood that forms in horizontal stems of trees to reestablish the tree's architecture; production of large amounts of storage parenchyma by the vascular cambium; concentric rings of vascular cambium as in beets; and differential activity of the vascular cambium producing large amounts of parenchyma in some areas and vascular tissue in others. Some arborescent monocots produce a vascular cambium outside the vascular bundles. The cambium produces an outer parenchymatous secondary cortex and an inner mass of procambial strands and lignified conjunctive tissue. (p. 374)
10. Products from wood include lumber, pulp and paper, fuel, charcoal, fabric, rope, sugars, spices, dyes, drugs, tanning fluids, chewing gum, and adhesives. (p. 374-377)

Constructing a Concept Map

Your map should incorporate headings from the chapter and show the relationships between main concepts and subconcepts.

Fill-in-the-Blanks

1. secondary
2. fusiform
3. xylem, phloem
4. summer wood (late wood)
6. annual rings or growth rings
7. tracheids, fibers
8. pores or vessels
9. sapwood
10. cross section
11. knot
12. specific gravity
13. reaction wood
14. bark
15. fibers
16. Lenticels
17. parenchyma
18. Pulp
19. Plywood
20. Charcoal

Putting Your Knowledge to Work

1. B. The "tree" is an arborescent monocot.
2. A. Rays in a tangential section appear as stacks of circles.
3. D. The wood has numerous resin canals lined with parenchyma cells.
4. D. Removal of the bark prevents movement of sugars from the leaves to the roots. Once stored sugars in the roots are used up they and the tree die.
5. C. Yes. The heart will be at exactly the same height as when he carved it.

Doing Botany Yourself

Callus tissue can be cultured from cortical or pith tissue in the stem. Buds that are not dormant could be grafted into the callus. A callus could be sampled and examined for the formation of vascular tissue below the buds. If vascular tissue does form, it would indicate that a substance or substances from the buds is stimulating vascular differentiation. The buds could then be placed on agar blocks and any exudates collected in the agar. These blocks could then be placed onto the callus in the same manner as the buds. If vascular tissue also differentiates with just the exudates from the buds, it would further support a role for the buds in differentiation of vascular tissue in the stem.

CHAPTER 17 REPRODUCTIVE MORPHOLOGY

After You Have Read the Chapter

1. What are the four parts of the flower?
2. Compare the pollen to microspores?
3. Describe the various forms of placentation for ovules in the carpel.
4. In what ways can the petals (corolla) vary within a flower?
5. What are thought to be the evolutionary origins of each part of floral parts?
6. Describe the formation of the egg in angiosperm megagametophytes.
7. Compare monocotyledons and dicotyledons.
8. What types of breeding systems are common in angiosperms?
9. What types of pollination and pollinator relationships have evolved in angiosperms?
10. Describe the types of fruits produced by angiosperms.
11. What types of environmental conditions regulate seed germination?
12. How are seeds and fruits dispersed?

Constructing a Concept Map

Using the instructions in the introduction to the Study Guide, construct a concept map of the ideas in the chapter.

Fill-in-the-Blanks

1. Angiosperms are easily identified by their distinctive reproductive structures called ____________________. (P. 384
2. Insect pollinated plants usually have colored ________________________ that attract the pollinators and guide through the flower for effective pollination. (p. 387)
3. Meiosis in the ____________________of the stamen produces microspores.(p. 390–391)
4. The ovary contains the ______________, which are attached to the carpels by placentae. (p. 385-386)
5. A ______________________ flower may contain a pistil but no petals. (p. 389)
6. When shed ______________________ of angiosperms usually contain two cells. (p. 391)
7. The ____________________ (female gametophyte) of most angiosperms develops from one megaspore, though many angiosperms deviate from this pattern. (p. 391-392)
8. The pollen tube enters the ovule through the ________________ of the ovule formed where the integuments meet. (p. 393)
9. One sperm in angiosperm fertilization fuses with the polar nuclei to form the _______________, which serves as a source of nutrition for the developing embryo. (p. 393)
10. One sperm fuses with the polar nuclei to form ________________________ that can provide food to the embryo or developing seedling. (p. 393)
11. Plants that produce perfect flowers but require another source of pollen to produce viable seeds have a ____________________ breeding system. (p. 395-396)
12. During _______________, pollen is transferred from the anther to the stigma. (p. 398-400)
13. Flowers pollinated by _______________ have honey guides visible under UV light. (p. 400)
14. Flowers pollinated by _______________ have tubular corollas and strong scents and often bloom at night. (p. 402-403)

15. In dry fruits the exocarp, mesocarp, and endocarp are fused into a thin _______. (p. 402-403)
16. Fruits often form as tissues divide and grow in response to ________________ secreted by the developing seed(s). (p. 403)
17. A seed will always contain a _______________, a food source, and an embryo. (p. 404)
18. Some seeds require that a thick seed coat be broken mechanically; this treatment, called ______________________, allows water and oxygen to enter the seed. (p. 405)
19. Thousands of seeds buried in the soil make a natural __________________ that can restore a disturbed habitat. (p. 405)
20. Seeds with plumes or wings are easily dispersed by ____________. (p. 407)

Putting Your Knowledge to Work

1. A plant ecologist is observing the behavior of pollinators that are visiting a flower. He expects to see hummingbird pollinators. Why?
   A. The flower is an iridescent white most visible at sunset and dawn.
   B. The flower has a strong, fruity scent.
   C. The flower consists of a regular corolla with separate petals.
   D. The flowers are red and have a large amount of nectar at the base of the corolla tube.

2. Anthropologists are studying a remote population. They have determined that a fungal infection wiped out all of a particular fruit tree in the surrounding area about five years ago. The natives have noticed that their major source of meat, a small rodent, is also disappearing. What could the anthropologists conclude?
   A. The decline in the availability of fruit has put increased pressure on the rodent as food.
   B. The rodents may have used the fruit tree as a major food source, and their disappearance has led to a decline in the rodent population.
   C. The fugus pathogen may have also infected the rodent population.
   D. The rodents may have emigrated out of the area to escape predation by the natives.

3. A flower that is incomplete
   A. is missing any floral whorl.
   B. is also inferior.
   C. always has petals, but not sepals.
   D. has pollen or ovules, but not both.

4. A plant anatomist examines a cross section of a seed. The seed has a large amount of endosperm, one cotyledon modified for absorption, and a plumule covered by a coleoptile. He concludes that the seed belongs to a type of
   A. dicotyledon with one cotyledon modified into a cover for the root.
   B. monocot such as corn or other grass.
   C. gymnosperm.
   D. seed that has undergone stratification.

5. A botanist finds a population of tambalacoque trees on an island 100 miles off the coast of India. Which of the following would be a viable explanation for the presence of the trees.
   A. A mutation has removed the requirement for scarification for germination of the trees.
   B. A period of submersion in seawater caused the seeds to swell, bursting the seed coats and allowing the seeds to germinate.
   C. Dodos introduced recently on the island have provided a natural source of scarification.
   D. The fruits were dropped by colonies of seagulls flying over the island, and the impact scarified the seeds.

Doing Botany Yourself

You are given the challenge of demonstrating that the strawberry ``fruit'' requires the achenes on its surface (often mistaken for seeds) in order to develop. How could you do this?

ANSWERS TO CHAPTER 17 STUDY GUIDE QUESTIONS

After You Have Read the Chapter

1. The four whorls of the flower are the sepals (calyx), petals (corolla), stamens (androecium) and carpels (gynoecium). (p. 384)
2. Microspores that undergo endosporic development into pollen grains. The microspore secretes a thick, two-layered cell wall. The microspore protoplasts divides to form a tube cell(in the germinated grain develops into the pollen tube to carry sperm to the egg) and a generative cell(in the germinated grain divides to form two sperm). The germinated pollen grain develops into the male gametophyte. (p. 390-391)
3. Ovules can have marginal placentation (along the margin of the suture), parietal placentation (on the ovary wall), free-central placentation (on a central stalk in a single carpel), or axial placentation (on the axis of a compound ovary). (p. 386)
4. Petals may vary in color from brightly colored to white, in number per flower from species to species, in arrangement (free or fused), and in symmetry (radial or bilateral), and may or may not have an elongated corolla tube. (p. 387–388)
5. The origins of flower parts vary. Sepals probably evolved from leaves. Petals probably evolved from aborted stamens, and in some flowers from sepals or leaves. Stamens may have evolved from fertile leaves or branching systems. Carpels may have evolved from folded or rolled fertile leaves or from structures similar to the ovule-bearing structures of some seed ferns. (p. 388–389)
6. The angiosperm egg does not develop within an archegonium. It is one of several cells in a specialized female gametophyte called an embryo sac. The embryo sac develops in the ovule. In most angiosperms one megaspore mother cell divides by meiosis to form four megaspores. One of these divides by mitosis to form eight free nuclei. These will organize themselves into the egg apparatus (two synergid cells and an egg cell) near the micropyle, two polar nuclei near the center, and three antipodal cells opposite the micropyle. (p. 391–392)
7. Monocots have one cotyledon, floral parts in threes, pollen with one furrow or pore, parallel leaf venation, complex arrangements of vascular bundles in the stem, and no secondary growth. They are mostly herbs. Dicots have two cotyledons, floral parts in multiples of four or five, pollen with three furrows or pores, netlike leaf venation, and vascular bundles in a ring in the stem; secondary growth is common. Many trees, shrubs, and herbs are dicots. (p. 393–394)
8. Angiosperms have many types of breeding systems. Each species hasevolved a particular pattern with special adaptations. Species may be monoecious or dioecious. Many display self-incompatibility while others are self-compatible. Many can reproduce vegetatively to form separate populations or large clones. (p. 395–398)
9. Angiosperm pollen is transferred from anther to stigma by a variety of agents including wind, water, and animal pollinators. Animal pollinators include insects such as butterflies, moths, wasps, flies, and bees; birds such as humming birds; and mammals such as bats, small rodents, and humans. Each species has evolved adaptations best suited to its particular pollination mechanism or agent. (p. 398-401)

10. Fruits may be fleshy with three layers—the skin (exocarp), the fleshy portion (mesocarp), and interior portion surrounding the seed (endocarp). The endocarp may be fleshy or hard. Dry fruits have a pericarp in which the three layers are fused into a single layer that is dry at maturity. Fruits may contain only the enlarged ovary and its seeds or the ovary plus other flower parts. A single fruit may be formed from a single ovary, from many ovaries in a single flower (aggregate fruit), or from many flowers (multiple fruit). (p. 402-403)
11. Environmental conditions must be correct to prepare seeds for germination. These conditions may involve a cold treatment (stratification), physical breakage of the seed coat (scarification), removal of inhibitors by exposure to light or leaching, or maintenance in the dark. When seeds are ready to germinate, the environment must provide the proper amount of water, the proper temperature, and adequate oxygen. (p. 404-405)
12. Seeds may be dispersed by wind, water, animals, or self-dispersal. (p. 407–409)

Constructing a Concept Map

Your map should incorporate headings from the chapter and show the relationships between main concepts and subconcepts.

Fill-in-the-Blanks

1. flowers
2. receptacle
3. anther, microspores (pollen grains)
4. ovules
5. gynoecium
6. vegetative, generative
7. embryo sac
8. micropyle
9. endosperm
10. Double fertilization
11. self-incompatible
12. pollination
13. bees
14. moths
15. pericarp
16. hormones
17. seed coat
18. scarification
19. seed bank
20. wind

Putting Your Knowledge to Work

1. D. The flowers are red and have a large amount of nectar at the base of the corolla tube.
2. B. The rodents may have used the fruit trees as a major food source, and their disappearance has led to a decline in the rodent population.
3. A. is missing any floral whorl.
4. B. monocot such as corn or grass.
5. A. A mutation has removed the requirement for scarification for germination of the seeds.

Doing Botany Yourself

Fruit-development is often triggered by auxin produced in fertilized seeds. The seeds of strawberries are contained in the achenes on the surface of the strawberry. If the achenes are producing auxin that is triggering enlargement and maturation of the strawberry tissue around the achenes, the removal of the achenes should stop the development of strawberry tissue.

Possible treatments include the following:

1. Removing all the achenes from a fruit
2. Removing seeds on one half of a fruit
3. Removing a small number of achenes from small areas of a fruit

If only the regions of the strawberry containing the achenes develop, then the achenes must be required for the underlying tissues to grow and ripen into the mature strawberry fruit.

CHAPTER 18 PLANT HORMONES

After You Have Read the Chapter

1. Why could Darwin be called the first plant physiologist?
2. How was the plant hormone auxin identified?
3. What evidence does not support the acid-hypothesis of auxin-induced growth?
4. Describe the roles of ethylene, IAA, and cytokinins in regulating apical dominance.
5. What types of developmental processes involve gibberellic acids?
6. Relate the discovery of gibberellic acids and cytokinins.
7. Why could Ca++ ions be considered to be second messengers in plant cells?
8. What observations led to the identification of ethylene as a plant hormone?
9. Compare the effects of ABA on plants to the effects of other plant hormones.
10. How do oligosaccharins and other plant hormones differ in their regulation of plant growth?

Constructing a Concept Map

Using the instructions in the introduction of the Study Guide, construct a concept map of the ideas in the chapter.

Fill-in-the-Blanks

1. The phototropic curvature of grass coleoptiles was shown by _________ to be regulated by the differential transport of _____ synthesized in the tip of the coleoptile. (p. 417, 421)
2. A mixture of synthetic auxins, ___________________, was used as a defoliant during the Viet Nam War to defoliate jungle vegetation. (p. 422)
3. Auxin is transported ______________________ in stems and ___________________ in roots. (p. 422-423
4. The ______________________ hypothesis proposes that IAA triggers the transport of H+ ions into the cell wall causing bonds between cellulose microfibrils to break and the cell to expand. (p. 423)
5. Falling levels of auxin in aging leaves or fruits trigger changes in the ____________________, resulting in causing leaf or fruit drop. (p. 424)
6. Undifferentiated tissue exposed to auxin and low levels of sucrose will differentiate into _______________. (p. 424)
7. Herring-sperm DNA, coconut milk, and yeast extracts have all been used as sources of _________________ to allow plant cells to be cultured. (p. 429-430)
8. Auxins stimulate the process of induction of_________________ _______________ on cuttings. (p. 424)
9. ________________________ fruits can develop in the absence of seeds if an adequate supply of auxin or gibberellic acid is present. (p. 424)
10. IAA can cause the release of ______________________ into the cytosol activating calmodulin that may cause physiological changes in cells. (p. 425)
11. Gibberellic acid stimulates the secretion of ___________________, an enzyme that converts starch to sugar, from the aleurone layer of grains to initiate germination. (p. 427)
12. Cytokinins stimulate ____________________________ by hastening the transition from the G_2 phase to mitosis of the cell cycle. (p. 431)
13. A big increase in respiration and oxygen consumption signals the onset of the _____________________ stage of fruit ripening and is initiated by rising ethylene in the fruit tissues. (p. 433)

14. Tissue changes in waterlogged plant organs and ______________ of stems are stimulated by a high concentration of ethylene. (p. 435)
15. The accumulation of ________________ during water stress causes leakage of ______________ from the guard cells, which causes the stomates to close. (p. 436)
16. Oligosaccharins are released by enzymes from the ________________ and regulate specific functions of the cell. (p. 437)
17. ______________________ are steroid hormones produced by many plant species that may influence light-regulated genes. (p. 427)
18. Gibberellic acids stimulate internode ________________ that causes the bolting that occurs at flowering of rosette plants. (p. 428)
19. The ____________ of flowers is influenced by ethylene and gibberellic acid. (p. 434)
20. Tissue levels of hormones vary with the rate of ______________, breakdown, or inactivation. (p. 431-432)

Putting Your Knowledge to Work

1. A plant physiologist finds that plants that grow on low-calcium medium have short internodes. Treatment of these plants with auxin or gibberellic acid does not reverse this inhibition. What does this indicate about the role of Ca^{++} ions in the action of plant hormones?
   A. The response of plant tissue to these hormones may require calcium ions.
   B. Auxin and GA stimulate the secretion of Ca^{+} ions into the cell wall causing increased growth.
   C. Calcium ions are not involved in the regulation of elongation growth.
   D. Endogenous auxin and gibberellins do not regulate cell elongation in growing tissue.

2. A greenhouse technician is given the job of producing bushy, many-flowered chrysanthemums demanded by customers. What would you suggest he do?
   A. Expose the young plants to ethylene.
   B. Treat the plants with GA to stimulate bolting required for flowering.
   C. Remove the apical buds on all stems to stimulate axillary buds.
   D. Treat the plants with ABA to stimulate the plants to produce more flower buds.

3. A chef is preparing for a large party. The pears he has ordered are not ripe enough to use. The party is in three days. What is the best way for him to hasten the ripening process?
   A. Place the pears in the dark.
   B. Place the pears in the refrigerator.
   C. Place the pears on the window sill.
   D. Place the pears in brown paper bags with ripe apples.

4. You are to propagate a new cultivar of elm trees by tissue culture. Which of the following would not be included in the culture medium?
   A. abscisic acid
   B. auxin
   C. cytokinin
   D. calcium

5. You want to induce bud formation in a tissue callus. What should you do?
   A. Increase the ratio of cytokinin to auxin in the medium.
   B. Make incisions in the callus to simulate wounding.
   C. Decrease the ratio of GA to auxin in the medium.
   D. Place the callus in light for 24 hours.

Doing Botany Yourself

Roots form on cuttings from woody plants on the end of the twig closest to the trunk even if the twigs are inverted. High levels of auxin are known to stimulate root formation. You think auxins in the twig are highest in the portion nearest the trunk and thus cause root formation. How could you prove this without any instrumentation to measure levels of auxin in tissue extracts?

ANSWERS TO CHAPTER 18 STUDY GUIDE QUESTIONS

After You Have Read the Chapter

1. Charles Darwin was the first to determine that a substance produced in the grass coleoptile regulated its phototropic reactions. He was the first to link a substance with a morphogenic response. This is one of the major areas of plant physiological study. (p. 417)
2. The plant hormone auxin was identified through experiments involving phototropic responses in coleoptiles. (p. 421)
3. IAA may induce growth under conditions in which acid does not. IAA-induced growth can be separated from wall acidification. IAA-induced growth and wall acidification are not causually linked.Acid-inuced growth stops after 1-3 hours; IAA-induced growth lasts much longer. IAA seems to cause gene induction; acid does not. (p. 423)
4. IAA from the tip may cause cells around the buds to produce ethylene that inhibits the growth of the buds. Cytokinins from the roots stimulate bud growth. Buds far enough from the tip to not be inhibited can be stimulated by cytokinins to develop. (p. 423-424)
5. Bakanae of rice seedlings; internode elongation in mature regions of trees, shrubs, and some grasses; greass seed germination;juvenility;flowering in rosette plants; and formation of "seedles" grapes. (p. 426-428)
6. Gibberellic acids were first islated by Japanese scientists from rice seedlings infected with the fungus that causes bakanae. Skoog and Miller isolated the first cytokinin, kinetin, from a herring-sperm DNA preparation. (p. 426–428)
7. Calcium ions are released into the cytosol by treatment of cells with growth-stimulating levels of auxin. The calcium ions activate calmodulin, a calcium-activated protein that can alter ion pumps and physiological processes in cells. Calcium also seems to be involved in the responses of cells to cytokinins. (p. 422, 432)
8. In Germany it was noted that plants growing near lamps powered by ``illuminating gas" had unusually thick and short stems and lost their leaves. It was later determined that the component of the gas causing these effects was ethylene. (p. 432)
9. ABA often inhibits the effects of other hormones. For example, it inhibits amylase action in seed germination, offsets cytokinin inhibition in senescence, and reduces wall elasticity promoted by IAA. (p. 436-437)
10. Each oligosaccharin is specific for a certain effect, whereas plant hormones often affect a wide variety of processes. Oligosaccharins stimulate specific enzymes; the actual effects of plant hormones are often more indirect, if even known. (p. 437)

Constructing a Concept Map

Your map should incorporate headings from the chapter and show the relationships between main concepts and subconcepts.

Fill-in-the-Blanks

1. Went, auxin
2. agent orange
3. basipetally, acropetally
4. acid-induced growth
5. abscission layer
6. xylem
7. cytokinins
8. adventitious roots
9. Parthenocarpic
10. calcium ions
11. alpha amylase
12. cell division
13. climacteric
14. epinasty
15. ABA, potassium ions
16. cell wall
17. Brassinosteroids
18. elongation
19. sex
20. synthesis

Putting Your Knowledge to Work

1. A. The response of plant tissue to these hormones may require calcium ions.
2. C. Remove the apical buds on all stems to stimulate axillary buds.
3. D. Place the pears in brown paper bags with ripe apples.
4. A. abscisic acid
5. A. Increase the ratio of cytokinin to auxin in the medium.

Doing Botany Yourself

Portions of the twigs could be excised from freshly cut twigs. Extracts could be made of these samples. These extracts could be applied in a bioassay. One bioassay would be the one used by Went in his examination of phototropism in the text. The extracts could be incorporated into agar blocks and applied to decapitated coleoptiles to cause curvature. Those agar blocks with the higher levels of auxin would cause greater curvature. If the lower ends of the twigs had the higher levels of auxin, the extracts from the twigs would cause greater curvature in the coleoptiles than those from the upper end.

CHAPTER 19 HOW PLANTS RESPOND TO ENVIRONMENTAL STIMULI

After You Have Read the Chapter

1. Define tropism and describe the types of environmental signals that can trigger tropisms?
2. What three hypotheses have been proposed to explain phototropic curvature in coleoptiles and stems?
3. How are auxin and Ca^{++} ions thought to affect the gravitropic response of roots?
4. What are the roles of hormones and light in thigmotropism?
5. How do seismonastic responses occur in plants as a result of shaking or contact stimuli?
6. What mechanism is involved in ``sleep movement," or nyctinasty?
7. How do plants time their flowering to maximize the chances of reproductive success?
8. Explain what is meant by a 10-hour photoperiod.
9. What is the role of phytochrome in plant development?
10. Describe the role of a biological clock in regulating a circadian rhythmn.

Constructing a Concept Map

Using the explanation provided in the introduction to the Study Guide, construct a concept map of the ideas in the chapter. Use the headings in the text as source material.

Fill-in-the-Blanks

1. A stem that bends away from the ground is said to be ________________ gravitropic. (p. 444)
2. Timothy Caspar's work questions the theory that roots sense changes in orientation to gravity by changes in the position of ____________________ in root cap cells. (p. 447)
3. In a root oriented horizontally IAA accumulates along the __________________ side and ________________ accumulate along the upper side of the root. (p. 450)
4. Tendrils and twining stems use ___________________________, a spiral growth pattern, toincreases the their chances of finding a support.(p. 450)
5. Seismonastic movements,leaf folding and leaf trap closure,result from a translation of a stimulus into an ______________________ signal that initiates events that result in a loss of turgor in cells that control leaf movements. (p. 451-453)
6. Sleep movements occur due to changes in __________________ in the cells of plant organs responsible for movement caused by ion movements into or out of these cells. (p. 445)
7. The short, stocky growth habit of plants growing outside is produced by the action of ____________________ genes such as the calmodulin gene and ethylene. (p. 454)
8. Garner and Allard were the first to show that differences in __________________ triggered flowering in some plants. (p. 454-456)
9. Flowering in long-night(short-day) and short-night(long-day) plants is regulated by the length of the ________________________. (p. 456)
10. After decades of searching, ___________________, the flowering hormone has still not been isolated. (p. 455)
11. The inactive form of phytochrome, P_r, absorbs ___________ light. (p. 458)
12. The dark conversion of ___________ allows plants to measure the length of the night. (p. 459)

13. Short-night(long-day) plants flower when the photoperiod is ________________ than the critical daylength. (p. 456)
14. A flash of far-red light during the dark period will ____________________ flowering in long-night plants. (p. 461)
15. Plants germinated and grown in the dark show signs of________________ including a lack of chlorophyll and elongated internodes. (p. 459)
16. Decreasing chlorophyll and mobilization of nutrients are part of the process of __________________ that accompany of aging and death of plant organs. (p. 461)
17. ______________________ is characterized by structural and chemical changes that bring on a period of decreased metabolism. (p. 461)
18. The release of tree buds from winter dormancy usually requires both a ____________ treatment and an increasing ________________. (p. 461-462)
19. The internal timing mechanism of circadian rhythms is called a ________________. (p. 462)
20. Many circadian rhythms are reset by ________________________ that controls the translation of mRNA's required to mediate the responses. (p. 463)

Putting Your Knowledge to Work

1. A sample of light-sensitive lettuce seeds are treated to 15-second bursts of red or far-red light in the following sequence: red, far-red, red, red, far-red, far-red, red. Which light treatment below would increase germination?
   A. far-red
   B. red
   C. far-red, red, far-red
   D. All of the above treatments would make the seeds ready for germination.

2. Which is true of a horizontally oriented root?
   A. Auxin accumulates on the upper side of the root.
   B. The levels of auxin in the amyloplasts rapidly decline in response to gravity.
   C. Calcium ions accumulate on the lower side of the root.
   D. Calcium ions are transported from the tip down the length of the root.

3. Which statement describes a photoperiod of 10 hours?
   A. It could consist of two-5 hour periods of light separated by 14 hours of darkness.
   B. It includes more than 10 hours of light.
   C. It includes 10 hours of light followed by 10 hours of dark.
   D. It includes 10 hours of light.

4. You work at a nursery. You are responsible for ensuring that the Christmas poinsettias flower in time for the holiday season. You know that the plants require more than 12 hours of darkness every 24 hours over a period of 6 weeks in order to flower. Which of the following would also be true?
   A. The plants are short-day plants.
   B. Turning on the lights in the greenhouse to check on the plants after 3 weeks of light treatments would put the plants 6 weeks behind schedule to flower.
   C. The plants are intermediate-day plants.
   D. The plants require 11 hours of light anytime in a 24-hour period.

5. A maple seedling in a pot is laid on its side until the stem bends 90°. It is then stood upright until the stem bends 90° again. It is laid down in the same original orientation until the stem bends 90° again. When it is stood upright again, what is the angle of the newest growth of the stem to the ground?
   A. 90 degrees
   B. 180 degrees
   C. 60 degrees
   D. 45 degrees

Doing Botany Yourself

Seedlings of trees that grow in the tropical rain forest often seek out a mature tree whose support will assist the tree in reaching the lighted regions of the canopy. Light is dim on the forest floor but sufficient to regulate phototropic responses. How would you show that the germinated seedling is growing toward the darkness of the space around a mature tree, not away from the brighter area between mature trees?

Answer Key to Chapter 19 Study Guide Questions

After You Have Read the Chapter

1. A tropism is plant growth toward or away from a stimulus. Many environmental stimuli can cause tropic movements including light, gravity, moisture, and touch. (p. 444)
2. The three hypotheses that address phototropic curvature in coleoptiles are (1) light destroys auxin on the lighted side of the coleoptile, slowing growth, (2) light causes auxin to move to the shaded side of the coleoptile, promoting growth, and (3) light decreases the sensitivity of tissue to auxin, slowing growth. (p. 444-445)
3. Auxin concentration in the responding tissues and calcium ions both affect the gravitropic responses of roots. (p. 450)
4. IAA and ethylene regulate the curvature of tendrils and stems that allows the organs to twine around supports. The curving growth requires light. When the support is touched in the dark the tendril or stem will not begin to change its growth pattern until it is in the light. The light requirement seems to be for ATP that is produced in the light. (p. 450-451)
5. Scismonastic responses require the rapid transmission of an electrical signal from sensitive cells to responding cells. (p. 451)
6. Nyctinastic movements are the result of changes in cell turgor in special responsive cells, sensitive to light and dark, which change the orientation of plant organs. (p. 453-454)
7. To maximize their chances of reproductive success, plants produce and distribute seeds that will germinate at the proper time of the year. Plants regulate reproduction by measuring day length with a biological clock that uses phytochrome. (p. 454-457)
8. A 10-hour photoperiod consists of 10 hours of light followed by 14 hours of dark. (p. 456)
9. Phytochrome is a pigment that senses the presence of light or the length of the light period in the environment of the plant. This allows the plant to change its growth pattern in response to its position in the habitat or to time reproduction to the seasons. (p. 456)
10. A circadian rhythm is a response that occurs on a nearly 24-hour cycle. The internal timing mechanism that controls the circadian rhythm is a biological clock. In plants the time-keeping mechanism of biological clocks seems to be able to be controlled by phytochrome. (p. 462-463)

Constructing a Concept Map

Compare your concept map to the sample in the introduction to the Study Guide.

Fill-in-the-Blanks

1. negatively
2. amyloplasts
3. lower, calcium ions
4. circumnutation
5. electrical
6. turgor

7. touch-induced
8. daylength, photoperiod
9. dark period
10. florigen
11. red
12. Pfr
13. longer
14. inhibits
15. etiolation
16. senesence
17. dormancy
18. cold, daylength
19. biological clock
20. phytochrome

Putting Your Knowledge to Work

1. B. red
2. C. Calcium ions accumulate on the lower side of the root.
3. B. It includes 10 hours of light.
4. B. Turning on the lights in the greenhouse to check on the plants after 3 weeks of light treatments would put the plants 6 weeks behind schedule to flower.
5. B. 180°

Doing Botany Yourself

Germinate the seedlings in five light-tight enclosures. Each container should have one small pinpoint hole either on the top or on one of the four sides.

If the seedling grows toward the pinhole of light, the seedling is positively phototropic.
If the seedling is negatively phototropic, it will grow 180° away from the pinhole into the darkest portion of the container.
If the seedling is positively dark-tropic, the seedling will grow most directly into the darkest portion of the container and not at any particular angle to the pinhole.

CHAPTER 20 SOILS AND PLANT NUTRITION

After You Have Read the Chapter

1. How is soil formed?
2. Differentiate between the A horizon, B horizon, and C horizon?
3. Compare a spodosol to an alfisol.
4. Describe the components of soil.
5. Compare silt to clay.
6. How does cation exchange affect the availability of nutrients in soil?
7. Why is humus an important component of soil?
8. After a heavy rain that saturates the soil describe the changes that will occur in water availability to plant roots over time?
9. What types of organisms contribute to the properties of soil?
10. What are three functions of essential elements?
11. Describe factors that affect the availability of nutrients to plants.

Constructing a Concept Map

Using the explanation provided in the introduction to the Study Guide, construct a concept map of the ideas in the chapter. Use the headings in the text as source material.

Fill-in-the-Blanks

1. Soils are the major source of ______________________ and ______________________ for plants. (p. 470)
2. The ______________________ in the soils of rain forests creates a rock-hard material when the rain forests are cut and the soil is exposed to the sun. (p. 470-471)
3. ______________________ soils are nearly equal mixtures of sand, silt, and clay. (p. 471)
4. Plants excrete H^+ ions which replace ______________________ on clay micelles in the soil, releasing these nutrients to the soil solution where they can be taken up by plant roots. (p. 472)
5. A 5-10-15 fertilizer will have ______________________ potassium than nitrogen. (p. 473)
6. Organic soils contain about 30% ______________________ which increases the water-holding and cation-exchange capacity of the soil. (p. 474)
7. An ______________________ is required for normal growth and reproduction and cannot be replaced by any other nutrient. (p. 477-478)
8. Lettuce grown ______________________ is grown without soil in a defined medium. (p. 477)
9. Plants growing near nuclear test sites accumulate ______________________ that can substitute for the element calcium in the plant's metabolism. (p. 482)
10. ______________________ adds calcium to the soil and removes H^+ ions, thus raising the soil pH. (p. 472)
11. Nitrates are converted by soil microorganisms to ______________ by the nitrate reductase system in the cells of the microbes. (p. 485)
12. Sulfur is absorbed by plants as ______________ which is reduced by ATP sufurylase. (p. 485)
13. Deficiency symptoms for a nonmobile element first show up in the ______________________ leaves. (p. 483)
14. To increase the ability to absorb water and nutrients a plant needs to increase its exposed ______________________. (p. 485)

15. Nitrification increases the nitrogen in the soil lost by ____________________. (p. 485)
16. In low-nutrient soils ____________________ help plants absorb nutrients more efficiently and to withstand environmental stresses better. (p. 485-486)
17. *Rhizobium* bacteria invade the roots of ____________________ to form nodules in which the bacteria fix atmospheric nitrogen. (p. 486)
18. Active and passive traps allow ____________________ plants to trap insects and small animals to provide nitrogen not available from nutrient-poor soils. (p. 488-489)
19. An ____________________ provides a xylem-to-xylem connection between the parasitic vascular plant and its host. (p. 490-491)
20. The ____________________ roots of many rain forest trees are an adaptation to low soil-nutrient levels in rain forest soil. (p. 491-492)

Putting Your Knowledge to Work

1. A geologist examines a soil sample from a murder site. He concludes that the corpse had been previously buried in a swamp before being dumped in bushes along the side of a superhighway. What features of the soil clinging to the corpse could lead the geologist to this conclusion?
   A. acid pH, high levels of organic material, and dark, almost black color
   B. alkaline pH and high clay content
   C. low pH and light color
   D. clumps of pebbles bound by sticky clay

2. Soil that has reached its field capacity has lost its
   A. gravitational water.
   B. capillary water.
   C. permanent wilting percentage.
   D. hygroscopic water.

3. A gardener notices that the leaves of her maple trees are turning yellow and are showing other features of nutrient deficiency. A soil test shows that the pH of the soil is 8.5. What soil treatment would you suggest she give the soil around the plants to reestablish a viable soil environment?
   A. Add large amounts of compost or another source of humus.
   B. Add a calcium phosphate fertilizer.
   C. Lime the soil.
   D. Add nitrogen-fixing symbionts to the soil.

4. You are raising tomato plants hydroponically. The plants grew well for a while, bbut suddenly the apical meristem of the stems have died. an examination of the root systems shows that the root tips are also brown and dead. You conclude that the level in the medium of which macronutrient is too low?
   A. sulfur
   B. phosphorous
   C. calcium
   D. nitrogen

5. Local teEnagers began showing up in emergency rooms of a small western town with all the symptoms of alcohol intoxication, but blood alcohol levels of 0.00. Once they recovered the local sheriff questioned them intensly. He found out that they all had smoked an herb cigarette sold by a local health food store. An analysis of the contents of the cigarette is likely to reveal the "tobacco" is really ground up:
   A. *Equisetum*.
   B. *Astragalus*.
   C. *Phacelia sericea*.
   D. *Sarracenia*.

Doing Botany Yourself

Develop a technique for determining the percentage of sand, silt, and clay in a soil sample.

Answer Key to Chapter 20 Study Guide Questions

After You Have Read the Chapter

1. Soil is formed by the pulverization of rock through the process of weathering caused by the action of wind, glaciers, and rain. (p. 470)
2. The A horizon is the topsoil. It lies just under the surface litter and extends 10-30 cm below the surface; has a pH near 7; contains 10-15% organic matter; and is dark in color. The B horizon is a region of the soil that is composed of medium-sized soil particles, lots of clay, little organic matter, and high levels of inorganic minerals. It lies between the A and C horizons about 30–60 cm below the soil surface. The C horizon occurs 90-120 cm below surface; contains unaltered rock fragments, mineral grains and lacks organic matter. (p. 470)
3. Spodosols are light-colored soils characteristic of wet, temperature regions. The A horizon has small amounts of organic matter and low pH. Alfisols are brown soils with more organic matter in the A horizon and are relatively fertile. They are common to the eastern United States. (p. 471)
4. Soil is rock that has been weathered into particles of less than 2 mm in diameter. In addition to weathered rock soil contains organic matter (humus), living organisms (bacteria, fungi, plants, and invertebrates), air, and water. (p. 471)
5. Water infiltration into silt is better than into clay. Clay holds water better than does silt. Clay is a better ion exchange medium than is silt. Silt has better aeration and workability than clay. Roots penetrate silt more easily than clay. (p. 472)
6. Clay micelles are negatively charged attracting cations. These cations can be displaced by other cations and released to the soil-water solution. H ions displace most other cations. Acid soils have most of their cations replaced by H ions. This causes nutrients to leech from the soil. Anions do not bind to soiled particles and are rapidly lost. This makes phosphorous and nitrogen levels in soils constantly limiting to plant growth. (p. 472)
7. Humus increases the water-holding capacity of the soil, the cation-exchange capacity of the soil, the aeration of the soil; and the nutrient reservoir of the soil. (p. 472-473)
8. Some water, gravitational water, will drain from the root zone. When all the gravitational water is drained that which remains is the field capacity. The water that is available for absorption by plants is cappilary water. Much of this can be lost by evaporation. When plants have absorbed all that they can the remaining soil water is the permanent wilting percentage. Further evaporation will leave the hygroscopic water that is held tightly by soil particles. (p. 475)
9. Earthworms aerate and process soil. Plants add organic matter to the soil when plant parts decompose and add carbon dioxide through root metabolism which acidifies the soil, releasing nutrients. Some plants actively secrete H ions and other substances into the soil, altering soil chemistry. Some plants secrete chemical substances that inhibit the growth of other plants. (p. 476)
10. Essential elements affect all plant processes including protein conformation, enzyme activity, osmotic changes in cells. They also function as coenzymes and essential parts of protein and pigment molecules and in the regulation of respiration. (p. 477)
11. Factors affecting the availability of soil nutrients for plant use include soil pH, the presence of chelators, the overall mineral composition of the soil, the ability of the plant to locate minerals in the soil, the surface area of the plant available for absorption, and the presence of organisms in the soil and associated with plant roots. (p. 482-488)

Constructing a Concept Map

Compare your concept map to the sample in the introduction to the Study Guide.

Fill-in-the-Blanks

1. water, nutrients
2. laterite
3. Loam
4. cations
5. more
6. humus
7. essential element
8. hydroponically
9. strontium
10. Liming
11. ammonia
12. sulfates
13. youngest
14. surface area
15. leaching
16. mycorrhizae
17. legumes
18. carnivorous
19. haustorium
20. agravitropic roots

Putting Your Knowledge to Work

1. A. acid pH, high levels of organic material and dark, almost black, color
2. A. hygroscopic water
3. C. Lime the soil.
4. C. phosphorous
5. B. *Astragalus.*

Doing Botany Yourself

Construct three pans with screens in the bottom. One with a screen 0.02-2mm diameter mesh; one with 0.02-0.002mm mesh; and one with slightly less than 0.002 mesh.

1. Take a soil sample that has been air dired. Gently break up the soil into a powder. Weigh the sample. Push the sample through the screen with the smallest mesh. Weigh the amount that goes through.
2. Push the remaining through the screen with the middle-sized mesh. Weigh the amount that goes through.
3. Push the remaining sample through the screen with the smallest mesh. Weigh the amount that goes through.
4. The soil that passed through the smallest mesh is the clay. Calculate its percentage by weight of the total sample.
5. The soil that passed through the medium mesh is the loam. Calculate its percentage by weight of the total sample.
6. The soil that the passed through the smallest mesh is the clay. Calculate its percentage by the weight of the total sample.

$$\text{Field capacity} = \frac{\text{Volume of water in lower container}}{\text{Volume of water added}} \times 100\%$$

CHAPTER 21 MOVEMENT OF WATER AND SOLUTES

After You Have Read the Chapter

1. Describe the factors that determine the water potential of a plant system.
2. What forces contribute to the movement of water through the xylem?
3. What evidence supports the transpiration-cohesion hypothesis for water movement through the xylem of a plant?
4. How have plants adapted to increase control of transpiration?
5. Explain the latest theory on the regulation of stomatal opening and closing.
6. Describe the adaptations plants have evolved as hydrophytes and xerophytes to deal with the water supply in their natural habitats.
7. Describe the structure and content of the sieve tube members.
8. Explain how sugars are moved through the phloem.
9. How are sugars loaded into and unloaded from the phloem?
10. How have aphids helped plant physiologists figure out how materials move in the phloem?

Constructing a Concept Map

Using the explanation in the introduction to the Study Guide, construct a concept map of the main ideas of the chapter. Use the headings in the text as source material.

Fill-in-the-Blanks

1. The loss of water through the stomates as water vapor is called ____________. (p. 498)
2. Conducting elements of the xylem include tracheids linked by bordered pits and ____________________. (p. 498-499)
3. When the __________________________ in root cells is less than that of the soil water will move into the root. (p. 500-501)
4. A fully turgid leaf cell will have a pressure potential ____________ than zero. (p. 501)
5. The water potential of any water-based solution is always ____________ than zero. (p. 500)
6. Water moves up small-diameter tubes by ____________ due to the adhesion of water molecules to the walls of the tubes. (p. 502-503)
7. Guttation is caused by a combination of high ____________ pressure, cool temperatures, and high atmospheric humidity. (p. 503)
8. The rate of transpiration from a leaf can be decreased by increasing the thickness of the ____________, a thin, moist layer adjacent to the leaf's surface. (p. 505)
9. CAM plants use ________________ water to fix carbon into sugars than do C4 or C3 plants. (p. 512
10. Plants under water stress produce high levels of the hormone ____________ which causes stomates to close, preventing further dessication. (p. 509)
11. The level of ____________ in guard cells best correlates with the pattern of opening and closing of stomates. (p. 509-510)
12. Guard cells ________________ at dawn as photosyntheis lowers the levels of carbon dioxide in the leaf tissue. (p. 510)

13. When guard cells are fully turgid when the stomates are fully ___________. (p. 509)
14. ___________ live under conditions of little water stress, poor gas exchange, and low light intensity. (p. 512)
15. Hartig proved that sugars are transported in the ___________ of the phloem. (p. 512)
16. Sieve plate pores in damaged phloem can be clogged with ______________________ a glucose polymer. (p. 514)
17. The Münch model of phloem transport states that a ___________ gradient provides the force for moving materials from source to sink. (p. 514-516)
18. Mature, photosynthesizing leaves are considered a ___________, a place from which sugars can be mobilized, while roots are usually considered a ___________, a place where sugars are rapidly metabolized or stored. (p. 515-516)
19. Phloem loading requires the assistance of ___________ whose cell walls have many plasmodesmata and invaginations like those of transfer cells. (p. 516)
20. Phloem loading requires ___________ to provide energy to establish a proton gradient across the sieve tube membrane which drives the cotransport of sucrose. (p. 516-517)
21. Plant physiologists have used ___________ to penetrate single sieve tubes to gain information on the content of phloem exudate and rates of transport. (p. 517-518)
22. More than 90% of the contents of the phloem are ___________, with the rest being amino acids, hormones, alkaloids, and inorganic ions. (p. 518)

Putting Your Knowledge to Work

1. Plants have developed many adaptations to maximize the benefits of available water. Which of the following is one of these adaptations?
   A. reorientation of leaves to increase leaf temperature
   B. decreasing the grams of water lost for each gram of carbon fixed
   C. increasing the leaf surface area
   D. growing more leaves during drought

2. Guard cells will open when:
   A. levels of ABA in the guard cells is increasing.
   B. K+ ions are pumped into the cell wall.
   C. The internal concentration of carbon dioxide in the cells is at its highest level.
   D. K+ ions are pumped into the cell.

3. If the pressure flow hypothesis of phloem transport is correct, then
   A. the sieve tube members must have intact plasma membranes to function.
   B. the water potential of the vessels must be greater than that of the sieve tubes.
   C. the sieve plate pores should be filled with callose in functional sieve tubes.
   D. ATP would not be required in order for transport to occur in the phloem.

4. A root cortex cell has a solute potential of −5 bars. The water potential of the soil is −3 bars. At what turgor pressure would the root no longer take up water?
   A. 0 bars
   B. −2 bars
   C. +2 bars
   D. +5 bars

5. If the transpiration-cohesion hypothesis of water transport in the xylem is correct, which of the following must also be true?
   A. The greatest rates of transport will occur when transpiration rates are greatest.
   B. Tree trunks will swell during the day as transpiration and transport increases.
   C. Cutting a stem will result in sap spurting out of the cut surfaces of the xylem.
   D. Leaf cells furthest from the site of transpiration have the lowest water potential.

Doing Botany Yourself

You are asked to determine the water potential of potato tubers that have been stored over the winter as a measure of viability. The higher the water potential in the tubers, the more turgid the tissue and presumably the better the seed stock they will make. You have no hygrometer or other special equipment. You have test tubes, dye, sucrose, water, and a scale. What can you do?

Answer Key to Chapter 21 Study Guide Questions

After You Have Read the Chapter

1. Water potential in a plant is determined by the pressure potential, solute potential (related to solute concentrations), and matric potential (related to wetability of exposed surfaces) in the various compartments of the system. (p. 500-501)
2. Capillarity, cohesion, tension, atmospheric pressure, root pressure, and evaporation all contribute to the movement of water from the roots to the leaves in the plant. (p. 502-503)
3. Evidences that support transpiration-cohesion hypothesis include: the water potential gradient favors movement of water from soil, through plant, to the atmosphere; water is under negative pressure in the xylem indicating it is being pulled by forces above, not pushed from below; the cohesion of the water in the column is greater than the forces pulling on it so it will not break; and the adhesion of the water molecules to the cell walls is strong enough to support a column of water as high as the tallest tree. (p. 505)
4. Plants alter rates of transpiration by physiological adaptations such as changing stomatal openings to increase water-use efficiency, leaf loss or dormancy during periods of high water stress, changing leaf position, circadian rhythms for stomatal opening and closure, ABA production during water stress to close stomates, and structural adaptations such as thick cuticles, sunken stomates, reduced leaf area, and stomatal number. (p. 505-507)
5. Stomatal opening and closing is regulated by two feedback loops. One is sensitive to internal carbon dioxide levels. The other is sensitive to rates of water loss from leaf tissues. (p. 510)
6. Hydrophytes have thin highly dissected leaves with much arenchyma tissue. This structure compensates for poor gas exchange and low light levels found in water environments. Xerophytes have a number of adaptations that compensate for high transpiration rates, high temperatures, and high light levels in their natural habitats. (p. 512)
7. Sieve tubes are formed from sieve tube elements stacked end-to-end. Each sieve tube element forms by the differentiation process of a single cell. Each cell was formed from the assymetric division of an initial that formed a small companion cell that provides energy for transport and the sieve tube element. The sieve tube element loses its nucleus and most organelles as the periphery of the cytosol fills with P-protein(slime) and callose(a glucose polymer). These materials clog sieve tubes that are severed or injured to prevent leakage. The end walls of each cell form sieve plates. The plates have numerous pores through which material moves between elements. (p. 513)
8. The pressure-flow model is the one currently used to explain transport through the phloem. A pressure gradient creates the force that moves the material through the phloem from source to sink where it is removed to maintain the gradient. Sugars are actively loaded into the phloem through a cotransport system driven by a proton gradient. (p. 514-515)
9. Sugars travel symplastically from chlorenchyma cells to the cell walls of the companion cells. Loading of sugars from the apolplast requires metabolic energy and is driven by a proton gradient. A H+ ion carrier pumps protons out of the sieve tube using ATP as K+ ions are being pumped in. Carrier transport of sugars is coupled to the diffusion of H+ ions back into the phloem. Unloading can be symplastic in actively growing sinks where a high gradient exists between phloem and surrounding tissue or apoplastic in other sinks. (p. 516)

10. Aphids can penetrate a single sieve tube with their mouth parts. The phloem exudate can be sampled as it is forced through the aphid body and out the back of the aphid as honeydew or the body can be separated from the head and the exudate collected directly. (p. 517-518)

Constructing a Concept Map

Compare your concept map to the sample in the introduction to the Study Guide.

Fill-in-the-Blanks

1. transpiration
2. vessels
3. water potential
4. greater
5. less than
6. capillarity
7. root
8. pulled
9. boundary layer
10. ABA
11. K^+ ions
12. open
13. open
14. Hydrophytes
15. sieve tube elements
16. callose
17. turgor-pressure
18. source, sink
19. companion cells
20. carbohydrates (sugars)
21. aphids
22. carbohydrates

Putting Your Knowledge to Work

1. B. decreasing the grams of water lost for each gram of carbon fixed
2. D. K+ ions are pumped into the cell.
3. A. the sieve tube members must have an intact plasma membranes to function.
4. C. +2 bars
5. A. The greatest rates of transport will occur when transpiration rates are greatest.

Doing Botany Yourself

When the tissues of a sample are at incipient plasmolysis, the pressure potential is equal to zero and the solute potential is equal to the water potential. If we can detect the point of incipient plasmolysis in a sample of potato tuber in a solution of known solute concentration, then we can determine the water potential of that tissue. This is the principle behind the Chardokov method for determining water potential.

The experiment would be set up as follows.

1. Set up two series of five test tubes. Make 50 ml of five solutions of sucrose. The series should be 2 M, 1.5 M, 1.0 M, 0.5 M and 0.25 M sucrose.
2. Pour 10 ml of each solution into each of two test tubes. Label each tube. Place the tubes in two rows beginning each with the 2 M tube and ending with the 0.25 M tube.
3. Put a drop of methylene blue dye in one series of tubes. Do not add dye to the other.

4. Cut 20 core sample of potato tissue from the same potato using a cork borer. Place them into a covered petri dish as each is cut to prevent water loss.
5. Put two cores into each of the tubes in the non-colored series of sucrose solutions.
6. Wait 45–60 minutes for the tissues to equilibrate with the sucrose solutions. Remove the pieces of potato.
7. Take a Pasteur pipet with a thin tip and draw up some colored solution from the 0.25 M tube, which is the mate of the uncolored tube. Put the tip of the pipet halfway down into the matching tube. Gently release a drop of colored sucrose solution (that is, release a drop from the 0.25 M colored sucrose tube into the 0.25 M clear sucrose tube).
8. If the drop sinks, water has been released from the tissue making the solution more dilute and less dense than that of the drop. The water potential of the tissue is greater than that of the solution.

   If the drop rises, water has been taken up by the tissue making the solution more dense than that of the drop, so the drop floats. The water potential of the tissue is less than the water potential of the solution.

   If the drop does not move, then water has not been released or taken up by the tissue; the water potential of the tissue is equal to the water potential of the solution. This is the point of incipient plasmolysis: the solute potential of the tissue is equal to the solute potential of the solution. For a solution in an open test tube the solute potential is equal to the water potential.

   Continue comparing the densities of the colored and noncolored paired solutions until you find the pair in which the drop remains stationary. This is the tube that contains a solution at the same water potential as the potato tissue sample.

CHAPTER 22 EVOLUTION

After You Have Read the Chapter

1. Compare the explanations for the diversity of life given by Plato, Aristotle, and Judeo-Christian tradition.
2. Contrast catastrophism and gradualism.
3. Compare the evolution theories of Lamarck and Darwin.
4. Name three observations made by Darwin on his voyage on the H.M.S. *Beagle* that influenced his theory of evolution.
5. What types of evidence did Darwin use to support his theory of evolution in *Origin of Species?*
6. Compare natural selection and artificial selection.
7. Explain the role of the population in the process of natural selection.
8. Compare the outcomes of the three types of natural selection.
9. What conditions must be met for the gene frequencies in a population to be maintained in a Hardy-Weinberg equilibrium?
10. How can genetic variation in a population be sustained over time?

Constructing a Concept Map

Using the explanation in the introduction to the Study Guide, construct a concept map of the main ideas in the chapter. Use the headings in the text as source material.

Fill-in-the-Blanks

1. The Greek philosophers and Judeo-Christian theology agree that species were ________________________ by an intelligent being. (p. 524)
2. Cuvier explained the abrupt changes in fossils in geological strata on the basis of a series of ___________________________ events such as volcanoes and earthquakes. (p. 525)
3. James Hutton's theory of __________________ suggested the slow change of geological features on the earth's surface and an ancient age for the earth. (p. 525)
4. Lamarck's ideas on the origin of species included the concept of ________________ to explain how offspring could be better adapted than their parents. (p. 527)
5. When ________________________ served as the naturalist on the H.M.S.Beagle his studies of the flora and fauna of S. America led to the development of his theory of evolution. (p. 527-528)
6. Darwin noted that finches on the ____________ arose from one mainland species through adaptation to various food sources. (p. 528)
7. Darwin used the ideas of Thomas Malthus on the impact of ____________________ growth onresources to develop his own theory of natural selection. (p. 529)
8. According to Darwin's theory of natural selection the amount of resources available is always ____________ than the amount of the resources needed by the population. (p. 530)
9. The theory of ____________ predicts that the best adapted individuals will leave the most surviving offspring. (p. 530)
10. According to Darwin the history of changes in a species is best represented by a ______________ not a straight line of descent. (p. 530)

11. ____________ provide physical evidence of change in a species over time through the preservation of recognizable structures. (p. 532)
12. The roots of *Azolla* are specialized leaves. These ``roots" show ____________ with true roots. (p. 533)
13. ________________ is the study of the geographical distribution of organisms. (p. 533)
14. Migration of a population to new environments followed by adaptation and speciation is called ________________________________. (p. 534)
15. ____________________ selection promotes the extreme variations of a phenotype. (p. 540)
16. The number of surviving offspring produced by an individual is a measure of its ____________. (p. 539, 542)
17. ____________________ can occur through migration and movement of pollen or seeds between populations. (p. 544)
18. Chance changes in the gene pool of a small population result in ______________. (p. 545)
19. ________________ accounts for the increased vigor of hybrids over their parents. (p. 546)
20. The macroevolution theory of __________________ predicts that the fossil record of a species will show periods of little change followed by a period of rapid change. (p. 547-548)

Putting Your Knowledge to Work

1. Assume skin color in pumpkins is regulated by a single gene so that orange skin is a dominant character and yellow skin is a recessive character. If the frequency of yellow pumpkins in a population is 0.09, predict how many pumpkins in a harvest of 1,000 pumpkins from that population will have orange skins.
   A. 910
   B. 91
   C. 490
   D. 700

2. The frequency of a recessive allele for hairy leaves is 0.5. If 1,000 offspring are produced for each of three generations, what will be the frequency of the hairy allele in the population after the third generation if all of a Hardy-Weinberg equilibrium is in effect?
   A. 1.0
   B. 0.5
   C. 0.25
   D. 0.1

3. A population of robins uses a type of berry as a food source. Cardinals with an average beak size larger than that of the robin is moving into the same habitat and is using the berry for food. Predict what effect the cardinals will have on the average size of a berry in the food plant population over several years.
   A. The cardinals will have no effect on berry size.
   B. The average size of a berry will increase.
   C. The average size of a berry will decrease.
   D. The berry plants will become extinct.

4. You have cucumber and watermelon plants in adjacent rows in your garden. When you harvest the watermelons, you notice they taste like cucumbers. The best explanation of this outcome is
   A. heterozygote superiority
   B. genetic drift.
   C. gene flow.
   D. self-incompatibility.

5. Yucca moth larvae feed only on yucca seeds. The yucca moth is the only pollinator of yucca. If herbicides sprayed on nearby fields kill the yucca moth, which of the following is not likely to happen?
   A. The yucca may become extinct.
   B. The number of offspring produced each year will increase.
   C. Other pollinators may adapt to pollinating yucca.
   D. Yucca may become self-fertilizing.

Doing Botany Yourself

A plant ecologist is studying *Baptisia* in Texas. He finds a lavender plant along a highway cut through an area of grazing land. He has never seen a lavender-flowered *Baptisia* in the region, but purple-flowered and white-flowered species are widely distributed. He believes the lavender-flowered species to be a hybrid between the other common forms. Suggest some types of evidence the plant ecologist would need to establish that the lavender-flowered form is a hybrid.

Answer Key to Chapter 22 Study Guide Questions

After You Have Read the Chapter

1. Plato and Aristotle believed that each species was created by a divine power. Plato believed that each species was a variation of an ideal form. Aristotle also wrote of the divine creation of eternal forms. Neither recognized that organisms evolve. Judeo-Christian tradition holds that each species was created individually during the six days of creation as described in Genesis . Each species is seen as perfectly suited to its role and this perfect match to be a reflection of the intelligence of God. Evolution of species is not part of this tradition. Human beings were also specially created to have dominion over nature and its inhabitants. (p. 524)
2. Catastrophism uses a series of destructive events to explain the natural history of the earth and the fossil record. Gradualism uses a series of slow but constant forces to explain the creation of mountain ranges and other geological formations such as sedimentary rocks containing fossils. This theory predicted a great age for the earth but gave no explanation for changes in species over time. (p. 525)
3. In Lamarck's theory, acquired characteristics induced by the environment are passed directly from one parent to its offspring, and this mechanism is the basis for the adaptation and change of species. Darwin's theory relies on inheritance of characters but does not give a mechanism for the passing on of characters. His theory locates change at the level of population and describes evolution through the process of natural selection. Variation occurs spontaneously and then individuals with these variations interact with the environment. (p. 527)
4. Some of the observations that influenced Darwin included the diverse flora and fauna of South America, the many different forms of finches and other animals on the Galápagos Islands and their similarity to mainland species, and the apparently perfect fit between animal form and their food. (p. 528)
5. Darwin used the following to support his theory: the great age of the earth, fossils, homology, convergence, biogeography, and artificial selection. (pp. 532)
6. Natural selection works at the population level while artificial selection can work at the level of the individual. Natural selection selects for the total phenotype, while artificial selection usually selects for a single trait or a few traits in the phenotype. (p. 536-538)
7. The members of a population interact with one another and the environment. Natural selection forces work on all the members of a population. Over time the frequencies of sets of genes in a population will change or not change in response to these forces, resulting in the maintenance of the original form or the increase in the frequency of another form. Individuals do not change, but the numbers of offspring a particular genetic type can leave behind can change. (p. 539)
8. Stabilizing natural selection maintains the average, or norm, and decreases the extremes. Directional selection selects for the extremes over the average. Diversifying selection occurs in rapidly changing environments to retard the average types and promote the extremes. (pp. 540)
9. The population must be large, with no migration or emigration, no mutations occurring, random reproduction, and no natural selection. (p. 542-544)
10. Genetic variation in a population is maintained over time by sexual reproduction, outbreeding, diploidy, heterozygote superiority, and heterosis. (pp. 545-546)

Constructing a Concept Map

Compare your concept map to the sample in the introduction to the Study Guide.

Fill-in-the-Blanks

1. specially created
2. catastrophic
3. gradualism
4. acquired characteristics
5. Charles Darwin
6. Galápagos Islands
7. population
8. exceed
9. natural selection
10. tree
11. Fossils
12. convergence
13. Biogeography
14. adaptive radiation
15. Directional
16. fitness
17. Gene flow
18. genetic drift
19. Heterosis
20. punctuated equilibrium

Putting Your Knowledge to Work

1. A. 910
2. B. 0.5
3. B. The average size of a berry will increase.
4. C. gene flow
5. B. The number of seeds offspring produced each year will increase.

Doing Botany Yourself

The plant ecologist could perform a variety of tests including examination of the numbers and forms of the chromosomes in the possible parents and compare these to those of the potential hybrid. He might expect to find a set from each potential parent and/or a polyploid condition based on a set from each parent.

He could perform biochemical analysis for pigments and other compounds that might be specific to each parent. If a mix of these appeared in the potential hybrid, it would be evidence for hybridization.

DNA and RNA sequence analysis that showed the presence of specific sequences found in each parent in the potential hybrid would be evidence for hybridization.

He could cross the parents and see if a hybrid similar to the lavender hybrid was produced. He could also cross this hybrid with the wild lavender hybrids to see if they are fertile. Infertility would not mean the lavender forms were not hybrids of the blue and white parents, however.

He could grow the wild hybrids in test plots to see if they grow under conditions intermediate to those of the potential parents. Hybrids often grow in intermediate habitats.

CHAPTER 23 SPECIATION

After You Have Read the Chapter

1. Describe the "species problem" that confronts plant biologists.
2. Compare the concepts of morphological and biological species.
3. What is a genetic species?
4. What is the role of reproductive isolation in the evolution of new species?
5. Compare prezygotic and postzygotic barriers to reproduction.
6. What are the major steps in the process of speciation?
7. Compare allopatric and parapatric speciation.
8. What role do ecotypes play in the evolution of new species?
9. What role does hybridization play in the evolution of plant species?
10. Explain how polyploidy creates new species.

Constructing a Concept Map

Using the explanation in the introduction to the Study Guide, construct a concept map of the main ideas in the chapter. Use the headings in the text as source material.

Fill-in-the-Blanks

1. The ____________ is the fundamental category of biological classification. (p. 554)
2. The classical concept of species is based on the idea of the ____________ species which is based on physical characteristics. (p. 554)
3. Many plants do not fit the concept of a ____________ species because morphologically distinct populations can interbreed freely if grown closely together. (p. 554)
4. The degree of genetic uniqueness of a species of sycamore found on the European continent from one on the Asian continent would be their ______________________________ from one another. (p. 556)
5. The concept of a ____________ species is important in the estimation of the degree of difference between a species and its closest relative. (p. 556)
6. Mechanisms that prevent the production of fertile hybrids are referred to as ____________. (p. 556)
7. Geographic barriers, microhabitats, and mechanical isolation are examples of ____________ barriers to reproduction. (p. 558)
8. A population whose flowers open at dawn would be ____________ isolated from members in the population whose flowers open only at sunset. (p. 558)
9. ____________ is a postzygotic barrier that results in infertile offspring when two species interbreed. (p. 559)
10. Gene frequencies in a population change because of natural selection, genetic drift, and ________________. (p. 560)
11. Populations of *Opuntia* cactus in New Mexico and Africa would be ____________ populations. (p. 560)
12. Cattails on the upper banks of a pond can undergo ____________ speciation with those growing near the water's edge. (p. 561)
13. Plants exhibit __________________________ allowing for the development of wide variations in size and form in response to environmental conditions. (p. 562)
14. A ____________ consists of plants that diverge morphologically along an environmental gradient. (p. 562)

15. ____________ differ genetically and physiologically from one another in response to local environmental conditions. (p. 563)
16. The offspring produced from the mating of two genetically distinct parents are ____________. (p. 564)
17. Many hybrids display ____________________allowing them to reproduce without producing gametes. (p. 564)
18. Alternate alleles can code for ______________________ that have distinct temperature optima or substrate affinities. (p. 565)
19. ____________________ can move genes between two populations around reproductive barriers when hybrids mate with parents. (p. 565)
20. Potatoes have a diploid number of 4N. This makes them ____________________. (p. 565-566)

Putting Your Knowledge to Work

1. A plant ecologist collects *Baptisia* specimens from several populations over a wide area. Members from the various populations interbreed when planted together in the university test plots. The resulting hybrids are fertile, but the second generation plants fail to set seed. This is an example of
   A. hybrid vigor.
   B. hybrid isolation.
   C. hybrid inviability.
   D. hybrid breakdown.

2. Fossil cones of the conifer genus *Auracaria* have one seed per ovuliferous scale, making the cones easily distinguishable from those of the genus *Pinus*. This paleospecies best fits the definition of
   A. an evolutionary species.
   B. a genetic species.
   C. a biological species.
   D. a morphological species.

3. A study of populations of tall blue stem grass along an east-west transect across Kansas shows that as one moves from east to west the populations become increasingly shorter and more drought tolerant. Hybrids can be found at the edges of adjacent populations.
   The best description of the populations along the transect would be:
   A. a cline.
   B. ecotypes.
   C. apomictic species.
   D. an introgression.

4. The first step in the evolution of a population into a new species is always
   A. a mutation in a gene that increases the survival of some members over that of others.
   B. the development of one or more reproductive barriers.
   C. geographical isolation.
   D. the development of polyploid offspring.

5. Two populations would not be considered distinct species if:
   A. artificial hybrids between the two populations are fully fertile.
   B. differences in body form were the result of morphological plasticity.
   C. the populations were growing in the same geographic region.
   D. gene flow occurs between the popula in their natural habitats.

Doing Botany Yourself

A plant ecologist sunflowers in the foot hills of the Sierra Madre Mountains. When he examines the specimens back in his lab he finds what he believes to be a new species. Describe the types of information he will need to establish that this is truly a new species of sunflowers.

Answer Key to Chapter 23 Study Guide Questions

After You Have Read the Chapter

1. The species problem for plant biologist is how to define a species because no one criterion seems to fit all functional species. Plant populations are constantly undergoing changes in response to environmental pressures and evolutionary processes. This means that populations, and species, are constantly changing making their definition difficult.(p. 554-555)
2. Morphological species is defined on the basis of physical characters. A biological species is defined on the basis of interpopulation fertility and the reproductive isolation of one population from another. (p. 554-555)
3. A genetic species is defined on the basis of its degree of genetic uniqueness in relation to other populations. This distance is defined by measurable differences in DNA fragments and gene sequences. (p. 556)
4. Reproductive isolation is the first step in the process of two populations becoming two distinct species. (p. 556-559)
5. Prezygotic barriers act before the formation of the zygote, preventing mating or the formation of a zygote. They include geographical, microhabitat, temporal, and mechanical barriers, and gametic isolation. Postzygotic barriers act after formation of the zygote and include hybrid inviability, hybrid sterility, and hybrid breakdown. (p. 558-559)
6. Variability within and between populations makes speciation possible. Reproductive barriers between populations accelerate speciation in concert with natural selection, genetic drift, and mutations. As interbreeding between populations declines, genetic drift continues, mutations accumulate, and selections acts, creating increasing genetic divergence and the development of two species. (p. 556-566)
7. Allopatric speciation occurs when two populations of the same species become geographically separated. The separated populations do not share the same range and accumulate genetic differences at different rates due to unequal selection. Parapatric speciation occurs when a species is distributed across an area with ranges that meet at a common border along an environmental feature. Plants at the border and in adjacent areas accumulate different genetic characters due to unequal selection pressure. (p. 560-562)
8. Ecotypes evolve from fragmented subpopulations distributed across a patchy or graded environment. Each subpopulation is exposed to different selection pressures. Ecotypes, however, can interbreed. Ecotypes can evolve into new species if reproductive barriers develop between the subpopulations. (p. 563)
9. Hybridization creates populations with unique genetic combinations. Hybrid populations can be more vigorous than either parent population; thus they are able to outcompete parental types. Hybrids allow for gene flow between hybrid and parent populations through introgression. The varieties that develop can each be adapted to slightly different habitats. (p. 564-565)
10. Hybrids have chromosomes from two different sources. In order to be fully fertile these chromosomes must be able to synapse and undergo meiosis. Polyploidy allows these hybrids to become fully fertile by creating homologous sets of chromosomes. The polyploid is now reproductively isolated from both parent populations and the nonpolyploid hybrids, creating an ``instant" species. (p. 565-566)

Constructing a Concept Map

Compare your concept map to the sample in the introduction to the Study Guide.

Fill-in-the-Blanks

1. species
2. morphological
3. biological
4. genetic distance
5. genetic
6. reproductive barriers
7. prezygotic
8. temporally
9. Hybrid sterility
10. mutation
11. allopatric
12. parapatric
13. morphological plasticity
14. cline
15. Ecotypes
16. hybrids
17. apomictically
18. allozymes
19. Introgression
20. polyploids

Putting Your Knowledge to Work

1. B. hybrid breakdown
2. D. a morphological species
3. B. ecotypes
4. B. the development of one or more reproductive barriers
5. D. gene flow occurs between the populations in their natural habitats

Doing Botany Yourself

To establish that he has found a new species he must establish that the population from which he collected the plant is morphologically distinct from other populations, has physiological and/or biochemically unique characters, has genetic distance from other populations, and does not normally reproduce with other populations in its native habitat. Comparisons need to be made with all other species of sunflowers native to the area using as many morphological characteristics as possible such as leaf size, shape, appearance(hairy, smooth, glandular,etc.) phyllotaxy; flower size, color, etc.;height; root systems, etc. If the plant is morphologically distinct, then any distinctive physiological adaptations or produces any unique enzymes or compounds could be additional evidence that the sunflower is a distinct species.

An examination of the chromosomes to look for chromosomes number and chromosome morphology would be important. Each species usually has a set chromosome number and unique shape and staining patterns of its chromosomes.

Hybridization studies with other species growing in the same habitat would provide information on the degree of reproductive isolation of the species as would a search for natural hybrids.

DNA hybridization, enzyme electrophoresis, DNA fingerprinting, gene sequencing, and RFLP's electrophoresis would all be important in assessing genetic distance between the new species and its closest relatives.

CHAPTER 24 SYSTEMS OF CLASSIFICATION

After You Have Read the Chapter

1. What was the motivation for the earliest studies of plants?
2. Describe the role played by Carolus Linnaeus in the development of the field of plant taxonomy.
3. The binomial designating American basswood is *Tilia americana.* What is the scientific name of American basswood?
4. Compare the criteria used to develop an artificial system of classification and a phylogenetic system.
5. Describe the information that is contained in a cladogram.
6. Since more than one cladogram can be constructed from the same data, how do taxonomists decide which is the most likely?
7. What five basic assumptions do taxonomists make when trying to develop a classification scheme?
8. What is a taxon?
9. Choose one character used to place organisms in one of the five kingdoms and describe the state of that character in members of each of the kingdoms.
10. What types of organisms have historically been classified by botanists in addition to plants?

Constructing a Concept Map

Using the explanation in the introduction to the Study Guide, construct a concept map of the main ideas in the chapter. Use the headings in the text as source material.

Fill-in-the-Blanks

1. The earliest publications for identifying plants focused on the ____________________value of plants. (p. 574)
2. John Ray was the first to divide flowering plants into ____________ and ____________. (p. 575)
3. Carolus Linnaeus developed a system for naming plants called the ____________ of nomenclature which identifies each plant by a unique two-part name. (p. 575)
4. *Rosa rugosa* is a rose that grows along the North American Atlantic coast. It belongs to the species ____________________. (p. 576)
5. Early __________________ systems of plant classification were meant to reflect the groups divinely created by God. (p. 578)
6. The first natural systems based on true phylogenetic relationships between plants were published by ____________ and ____________ between 1887 and 1910. (p. 580)
7. Phylogenetic classification systems stress ____________ as opposed to derived features of plants. (p. 580-581)
8. Charles Bessey rejected Engler and Prantl's idea that the ____________________ flower represented the primitive angiosperm flower. (p. 580-581)
9. Even when taxonomist use complex systems of data to construct classification systems they also must make __________________ as to the significance of each character. (p. 582)
10. When a ____________ is constructed, the character states used are given fixed values. (p. 583)
11. In a cladogram each fork represents a point of __________________________ one species from another ancestral species. (p. 584)
12. Systemasists use the principle of ______________________________ to choose the most probable cladogram from all those possible. (p. 553)

13. A genus consists of a group of closely related ______________________ that share a set of common features. (p. 575)
14. According to the theory of biological classification the greater the similarity between organisms the closer their ____________________________ relationships. (p. 579-580)
15. According to table 24.1 barley belongs to the Family ____________. (p. 586)
16. All ________________________ are classified in the Kingdom Monera. (p. 587)
17. Organisms with cells having membrane-bound nuclei, haploid life cycles, and autotrophic nutrition would be included in the Kingdom ____________. (p. 586)
18. Members of the Kingdom Plantae are eukaryotes with _______________ nutrition, cellulose cell walls, and _____________________ life cycle. (p. 587)
19. A classification system based on nucleotide sequences in DNA would be an example of a ________________. (p. 588)
20. According to the cladogram in figure 24.14 fungi are more closely related to ____________ than to plants. (p. 588)

Putting Your Knowledge to Work

1. A student developed a classification scheme for angiosperm families based on the number of petals in the flower. He chose petal number because the character is stable across a plant group, is easy to determine, and allows for quick identification of each plant group. This system is an example of
   A. a phylogenetic system.
   B. a system that will reveal evolutionary relationships between groups.
   C. a system that will not work because of the morphological plasticity of petal number in flowers.
   D. an artificial system.

2. A phycologist discovers a unicellular organism. It has anaerobic metabolism, flagellated cells, and reproduces by spores. This organism is most likely a member of the Kingdom
   A. Fungi.
   B. Animalia.
   C. Protoctista.
   D. Monera.

3. Using the cladogram in figure 24.12, choose the true statement about the evolution of monocots and dicots.
   A. One ancestor gave rise to both monocots and dicots.
   B. One ancestor gave rise to dicots, and a separate ancestor gave rise to all monocots.
   C. Monocots and dicots have no common ancestors.
   D. One ancestor gave rise to dicots A and dicots B, and a separate ancestor gave rise to monocots.

4. Two plants that are in the same order will not also be in the same
   A. species.
   B. genus
   C. class
   D. family.

Doing botany Yourself

You are given the task of developing a plant key to be used by botany students to identify the plants that grow around the science building on your campus. Describe the types of characters you could use in developing the key.

Answer Key to Chapter 24 Study Guide Questions

After You Have Read the Chapter

1. The earliest studies of plants were done to determine their usefulness in treating illness. (p. 574)
2. Carolus Linnaeus was the first to develop a system for naming individual plants for the purposes of classification. His binomial system of classification is the basis for all modern systems of classification. (p. 575)
3. The scientific name, *Tilia americana,* is the same as its binomial. (p. 575)
4. Artificial classification systems are constructed for purposes of identification only and are based on a few, easily recognized characters. They do not necessarily reflect evolutionary relationships between groups. Phylogenetic systems attempt to reflect evolutionary relationships between groups. The systems are constructed using ancestral and derived traits. (pp. 576-582)
5. Cladograms depict shared-derived and shared-primitive characters. They show direct relationships of ancestory and descent. (p. 582-584)
6. Taxonomists rely on the principle of parsimony that one should not make more than the minimum number of assumptions to explain any phenomenon. The cladogram that requires the fewest evolutionary changes is the one assumed to be the correct one. (p. 584)
7. The assumptions are listed on page 585.
8. A taxon is any level in a classification system such as kingdom, division, class, order, family, genus, or species. (p. 585)
9. Refer to table 24.2 for a list of characters. An example would be the character ``cell type." In Kingdom Monera the character appears as "prokaryotic"' in the other four kingdoms the characters appear as "eukaryotic". (p. 586)
10. At one time or another any organism not classified as an animal has been classified as a plant including viruses, bacteria, algae, dinoflagellates, euglenoids, and fungi. (p. 588)

Constructing a Concept Map

Compare your concept map to the sample in the introduction to the Study Guide.

Fill-in-the-Blanks

1. medicinal
2. monocotyledons, dicotyledons
3. binomial system
4. *Rosa rugosa*
5. natural
6. Engler, Prantl
7. primitive
8. *Magnolia*
9. Assumptions
10. cladogram
11. shared-derived
12. parsimony
13. species
14. evolutionary
15. Poaceae
16. prokaryotes
17. Protoctista
18. autotrophic, alternation of generations
19. molecular phylogeny
20. animals

Putting Your Knowledge to Work

1. D. an artificial system
2. A. Fungi
3. A. One ancestor gave rise to both monocots and dicots.
4. C. class

Doing Botany Yourself

Morphological characters are best for developing a key. Dichotomous keys are usually the easiest to use. These are keys in which one must choose which of two forms of the character the specimen one is identifying has. Then one moves to the choice associated with that character.

Most campuses have landscaping that includes ferns, gymnosperms, and angiosperms. A key should begin with choices that separate these three groups. Choices could include growth habit, branching, leaf type, leaf arrangement, leaf number per stem, cone features, flower structure(number of parts, color of petals, whether parts are fused or free),fruits and seeds,
features of twigs, types of hairs or scales, etc.

For example:

1. Produces flowers ....2.
1. Does not produce flowers...................................3.

2. Produces flowers with parts in threes................Monocotyledons
2. Produces flowers with parts in fours and fives.......Dicotyledons

3. Produces cones.......................................Gymnosperm
3. Produces spores.......................................Ferns

CHAPTER 25 BACTERIA

After You Have Read the Chapter

1. Compare bacterial cells to plant cells.
2. In what ways can bacterial cells undergo genetic recombination even though they do not have sexual reproduction?
3. Under what conditions can bacteria grow and reproduce?
4. Explain the current classification for bacteria.
5. In what ways are Archaebacteria like eukaryotes?
6. Compare the cyanobacteria and the chloroxybacteria.
7. What products are made from bacteria?
8. Describe the structure of viruses.
9. How are viruses classified?
10. In what ways are viruses unlike living organisms?

Constructing a Concept Map

Using the explanation in the introduction to the Study Guide, construct a concept map of the main ideas of the chapter. Use the headings in the text as source material.

Fill-in-the-Blanks

1. All bacteria are ____________________, lacking true nuclei or membrane-bound organelles. (p. 594)
2. Bacteria have cell walls composed of ________________________, a carbohydrate polymer interconnected with chains of amino acids. (p. 594)
3. The ___________________________ form upright, multicellular fruiting bodies. (p. 594)
4. The wall structure and/or lipid coating of many Gram-________________________ bacteria make them pathogens. (p. 595)
5. Bacteria play key roles in the ________________________ cycle that provides usable forms of nitrogen to plants. (p. 595)
6. Bacteria usually reproduce by ______________________ during which the genophore is duplicated. (p. 597)
7. Plasmids can move from one bacterial cell to another through ________________________. (p. 597)
8. As much a 90% of the carbon dioxide produced by decomposition is produced by__________________ bacteria. (p. 598)
9. The ___________________________ have unique features that are more like those of eukaryotes than other bacteria. (p. 599)
10. Phycobillins and chlorophyll a form the photosynthetic pigments of the ______________________. (p. 601)
11. Cyanobacteria living as symbionts often function as ________________________. (p. 601)
12. The bacterium ____________________________ is cultivated for human consumption in Mexico and is sold in health food stores in the US. (p. 603)
13. Species of _______________________ are used for insect control and production of ethanol for fuel. (p.604)
14. The production of yogurt and cheeses uses various species of __________________________. (p. 604)
15. The ___________________________ bacteria are sources of a number of antibiotics. (p. 604)
16. Municipal water supplies are fouled by toxins and odors produced by blooms of _________________. (p. 604)

17. Bacteria in the genus ______________________ produce highly poisonous toxins and heat resistant spores that make them extremely dangerous to humans. (p. 604)
18. Tobacco mosaic virus has a core formed from ______________________ and an outer protein coat. (p. 604)
19. Many viruses may be more genetically similar to their ______________ than to other viruses. (p. 607)
20. ______________________ are RNA molecules located inside the coats of viruses that are required for the replication of the virus. (p. 607)

Putting Your Knowledge to Work

1. A scientist has developed viruses that can kill any cyanobacterium. He created What will not be affected if he releases these into the environment?
   A. productivity of tropical ecosystems
   B. oxygen levels in the atmosphere
   C. nitrogen availability in the biosphere
   D. decomposition of dead organisms

2. If resistance to penicillin is carried on a plasmid, a small circle of DNA, in a bacterial cell, which of the following processes would not be a possible mechanism for the transfer of the plasmid between bacterial cells?
   A. transduction
   B. passage through a conjugation pilus
   C. transformation
   D. protoplast fusion during sexual reproduction

3. A terrorist wants to kill a large portion of the population of N Y City in a few days. Which would have the best chance of success?
   A. Put botulism toxin in the outflow filters of the water treatment plant.
   B. Distribute canned food to which endospores of toxic bacteria had been added before processing.
   C. Fertilize the water supply reservoirs to stimulate a bloom of cyanobacteria.
   D. Add *Bacillus thuringiensis* to bread products made at local bakeries.

4. A new disease has been discovered in a community in the Sonoma Valley of California. Cultures produced from the blood of those who have contracted the disease have produced naked pieces of RNA. When injected into mice these pieces of RNA cause similar symptoms to those seen in humans with the disease. The disease agent is most probably:
   A. a bacterium.
   B. a viroid.
   C. a virus.
   D. a virusoid.

Doing Botany Yourself

Soil bacteria often produce antibiotics to control the growth of competing bacteria. Develop an experimental protocol to look for antibiotics that are effective against human pathogenic bacteria.

Answer Key to Chapter 25 Study Guide Questions

After You Have Read the Chapter

1. Bacteria are prokaryotes and plants are eukaryotes. Bacterial cells have peptidoglycan cell walls and may also have a lipid covering around the wall; plant cells have cellulose cell walls that contain hemicelluloses, pectin, and protein. Bacteria have no mitochondria; plant cells do. Bacteria have a genophore containing their DNA; plant cells have chromosomes with nucleosomes as the structural subunit and a membrane-bound nucleus; bacteria divide by fission, plant cells divide by mitosis with phragmoplastic cytokinesis. Bacteria have flagella and/or cilia formed from flagellin; motile plant cells have eukaryotic flagella formed from microtublues with

9+2 structure. Photosynthetic plant cells have chloroplasts containing photosynthetic membranes, photosynthetic bacteria have photosynthetic membranes, but no chloroplasts. (p. 594-596)

2. Bacteria can take in DNA released into the environment or directly to other bacterial cells through conjugation, transformation, or transduction. This DNA can provide new functional genes to the cell. (p. 597)
3. Bacteria can live almost anywhere including the intestines of animals, the soil, clouds, water, and airborne particles. Bacteria can tolerate hot acids, freezing temperatures, and boiling hot springs, as well as crushing pressures at the bottom of the ocean. Some can form symbiotic relationships with plant roots. (p. 598)
4. The current system divides the Kingdom Monera into two subkingdoms-Eubcteria and Archaebacteria. Microbiologists include a third, the Eukaryota. Bergey's Manual of systemic Bacteriology, the most widely used reference, gives four divisions and seven classes. the manual does not always group by phylogeny, but often uses arbitrary characters like nonoxygenic photosynthesis. (P. 599-660)
5. Archaebacteria share the following traits with eukaryotes: glycoprotein and polysaccharide cell walls rather than peptidoglycan, DNA-dependent RNA polymerases, genes with introns. (p. 600-601)
6. Both are prokaryotes, both have chlorophyll *a,* and both release oxygen during photosynthesis. The cyanobacteria have phycobilins as accessory pigments in their photosynthetic systems, lack stacked thylakoids, and produce heterocysts for nitrogen-fixation and akinetes for reproduction. Chloroxybacteria have chlorophyll *b,* carotenoids, and stacked thylakoids as in green algae and plants. (p. 601)
7. Bacteria are used as food products, as insect-control agents, in the manufacture of fuel-grade alcohol, in the production of yogurt, cheese, and other dairy products, the production of wine, beer, sourdough bread, and sauerkraut, and the production of antibiotics, and in composting organic wastes. (p. 603-604)
8. Viruses lack any elements of cellular structure. They consist of a core of DNA or RNA surrounded by a protein coat. the protein coat may be attached to more complex structures. Most plant viruses have an RNA core and a coat composed of a few proteins. The genome is usually small as in TMV that contains only four genes. (p. 604-605)
9. Features such as host, tissue type they infect, type of nucleic acid, size, shape, nature of coat protein, and number of nucleic acid molecules in the core are all used to classify viruses. Animal and bacterial viruses are grouped together on the basis of their serotype. The serotype of a virus is determined by which antibody will react to an antigen(protein) on the surface of the virus. Plant viruses are classified using similar features, but are named on the basis of a prototype. This can separate closely-related animal and plant viruses into separate genera. (p. 605-607)
10. Viruses are unlike living organisms in that they cannot reproduce without the assistance of a living cell, they have no cellular structures, they do not increase in size or divide, and they do not respond to external stimuli. (p. 604-607)

Constructing a Concept Map

Compare your concept map to the sample in the introduction to the Study Guide.

Fill-in-the-Blanks

1. prokaryotes
2. peptidoglycan
3. myxobacteria
4. negative
5. nitrogen
6. fission
7. conjugation pili
8. saprobic
9. Archaebacteria
10. cyanobacteria
11. chloroplasts
12. *Spirillum*
13. *Bacillus*
14. *Lactobacillus*
15. actinomycetes
16. cyanobacteria
17. *Clostridium*
18. RNA
19. hosts
20. virusoids

Putting Your Knowledge to Work

1. D. decomposition of dead organisms
2. D. protoplast fusion during sexual reproduction
3. A. Put botulism toxin in the outflow filters of the water treatment plant.
4. B. a viroid.

Doing Botany Yourself

Collect soil samples. Place a small amount on a wire loop. Spread the soil particles apart with the sterile loop. Observe the types of bacteria colonies that grow on the plates. Remove each colony to a plate containing agar made with complete medium and a bacterium normally sensitive to antibiotics. Observe the colonies and the surrounding agar for several days. Colonies with clear areas around them are producing substances that are limiting the growth of the bacteria in the agar or is killing them. These substances could be antibiotics. The colony can be grown to a large volume so the substances can be extracted and analyzed.

CHAPTER 26 FUNGI

After You Have Read the Chapter

1. Describe the unique set of characters shared by the fungi.
2. Give a generalized life cycle of a fungus.
3. Distinguish the Zygomycetes from other fungi.
4. Describe the formation of ascospores.
5. What are some of the unusual characteristics of lichens?
6. What are the unique features of the Basidiomycetes?
7. Contrast the two forms of mycorrhizae.
8. What kinds of products are made using fungi?
9. How do plant pathogens impact humans?
10. What characteristics have been used to separate the slime molds and water molds from the true fungi?

Constructing a Concept Map

Using the explanation in the introduction to the Study Guide, construct a concept map of the main ideas in the chapter. Use the headings in the text as source material.

Fill-in-the-Blanks

1. The fungal thallus or mycelium is formed from thread-like ______________________. (p. 612)
2. A fungal hypha with two genetically distinct nuclei is ______________________. (p. 612)
3. Fungi store carbohydrate as ______________________ as do animals. (p. 612)
4. When two zygomycete gametangia fuse a ______________________ is formed. (p. 617)
5. Truffles, morels, and some bread molds are all ______________________. (p. 617-618)
6. To make beer, wine, and raised baked goods unicellular ascomycetes called ______________________. are required. (p. 617)
7. The ______________________ of the ascomycete ascogonium transfers nuclei to the antheridium of the opposite mating strain. (p. 618)
8. An ascoma produces many ______________________ in which ascospores are produced by meiosis. (p. 618)
9. Basidiomycetes and ascomycetes produce asexual ______________________ from the tips of specialized hyphae. (p. 622)
10. Deuteromycetes are fungi that lack a known ______________________ phase. (p. 618)
11. The thallus of a ______________________ is formed by a fungus and one or more algae living in a symbiotic relationship. (p. 620)
12. The ______________________ is a basidiomycete basidiomatum consisting of a cap and a stalk. (p. 622)
13. The dikaryotic condition of the basidiomycete hyphae is maintained by the formation of ______________________ during cell division. (p. 624)
14. The ______________________ of mushrooms bear the basidia that form basidiospores.
15. Spontaneous abortions, gangrene of the limbs, and hallucinations can result from ingesting drugs produced by grain infected with the ______________________ fungus. (p. 623)
16. The ______________________ are probably the closest relatives to the ancestor of the fungi. (p. 626)
17. The presence of ______________________ in the life cycles of ascomycetes and basidiomycetes is evidence of their close evolutionary relationship between the two groups. (p. 627)

18. Fungi are probably more closely related to ________________ than to plants. (p. 627)
19. ________________ have phagocytic nutrition and mobile cells unlike true fungi. (p. 627)
20. Water molds have flagellated cells and cell walls that contain ________________________ in addition to chitin. (p. 627)

Putting Your Knowledge to Work

1. What may be a direct result of the disappearance of forest mushroom species as a result of air pollution?
   A. the closing of gourmet restaurants
   B. an increased rate of die of forest trees
   C. an decrease in the level of soil nutrients in forests
   D. an increase in the number and diversity of lichens

2. A mycologist discovered some sporangia on a rotted log. The sporangium consisted of a stalk terminating in a sphere containing many black spores. She germinated some spores from the sporangium. The spores developed into cells with two whiplash flagella. She concluded that the sporangia were most likely produced by a:
   A. zygomycete.
   B. cellular slime mold.
   C. water mold.
   D. plasmodial slime mold.

3. A new fungus has been discovered, but it has never been known to reproduce sexually. A mycologist
examined the fungal hyphae and pronounced it to be a basidiomycete. Without basidiospores, how could he be sure?
   A. The hyphae were septate.
   B. The hyphae were coenocytic.
   C. The hyphae had clamp connections.
   D. Conidia were present on the tips of some hyphae

Doing Botany Yourself

You have been selected by your local mycology club to develop a key to recognize the local mushrooms. Explain the types of characters that would be important in identifying mushrooms.

ANSWER KEY TO CHAPTER 26 STUDY GUIDE QUESTIONS

After You Read the Chapter

1. Fungal structure is based on tubular, threads called hyphae that from mycelia, fruiting bodies, and spores. The hyphae may be aseptate or septate with each compartment being uninucleate, dikaryotic, or coenocytic. Fungi store carbohydrate as glycogen, have chitin cell walls, mitosis with a persistent nuclear membrane, reproduce only by spores, and produce no flagellated cells. (p. 612)
2. Fungi can reproduce asexually by the production of spores either at the tips of specialized hyphae(conidiophores) or in sporangia or by fragmentation of the mycelium. Fungi also produce sexual spores. These spores are produced when two haploid hyphae fuse producing a dikaryotic hypha. Fertilization occurs in the dikaryotic hypha forming zygotes that undergo meiosis to produce spores. Sexual spores are usually produced in a specialized fruiting body. (p. 612)

3. Zygomycete hyphae are aseptate(all other fungi have septate hyphae). The hyphae are coenocytic with no dikaryotic phase. Sexual reproduction occurs by the fusion of two gametangia(specialized hyphae produced on the growing thallus). The resulting zygosporangium that germinates hyphae that develop into sporangia. No fruiting bodies are formed. (p. 614)
4. The formation of ascospores requires the development of asci. On mating type develop an ascogomium. the ascogonium sprouts thin threads called trichogynes. The trichogyne grows toward and fuses with an antheridium produced on a mycelium of an opposite mating type. Nuclei from the antheridium pass through the trichogyne into the ascogonium. The nuclei from each hypha intermingle and pair up. The paired nuclei pass into ascogenous hyphae that grow out of the ascogonium. These become asci in which meiosis and mitosis produce eight ascospores. (p. 617)
5. Lichen fungi do not grow outside of lichens in nature. The lichen thallus is a unique form that occurs only when the fungus and alga grow together. They grow very slowly and can live for thousands of years. Lichens can withstand extreme environmental conditions of cold, heat, and drought. However, they are very sensitive to air pollution and disappear from urban areas where pollution is high. (p. 620)
6. The unique features of Basidiomycetes include the formation of clamp connections in their dikaryotic hyphae. All form basidia that produce basidiospores on their tips. Most form basidiomata that bear the basidia. (p. 620)
7. Endomycorrhizae are usually zygomycetes and occur in over 80% of all vascular plants. These produce hyphae, vesicles or arbuscles, that penetrate plant cells to facilitate nutrient transfer between fungus and plant. This type is most frequent in the tropics. Ectomycorrhizae are usually associated with trees and shrubs in temperate regions. they form a mantle over the root surface. They serve as root hairs to assist in the absorption of soil nutrients. They do not penetrate root tissue. Most are basidiomycetes. Some are ascomycetes. (p. 623-625)
8. Fungi are commercial sources of gallic acid used in photographic developers, dyes and ink. They are used to produce artificial flavorings, perfumes, chlorine, alcohol, plastics, toothpastes, soap, and to silver mirrors. (p. 625-627)
9. Rust fungi cause billions of dollars of lost revenue from crops and timber. Ergot infection of grain can cause a variety of medical disorders, stillbirths, and hallucinations. It also can be used to treat bleeding after childbirth, migraine headaches, heart disease, nervous stomach, and menopausal symptoms. (p. 626)
10. The presence of motile cells(amoeboid or flagellated) is the major feature separating slime molds and water molds from the true fungi. Chitin in the cell walls is lacking in some groups or is a minor component. Slime molds have phagocytic nutrition, not absorptive as in the fungi. the life cycles and reproduction of the slime molds and water molds is distinctive and quite different than any of the true fungi. (p. 627)

Constructing a Concept Map

Your map should incorporate headings from the chapter and show the relationships between main concepts and subconcepts.

Fill-in-the-blanks

1. hyphae
2. dikaryotic
3. glycogen
4. zygosporangium
5. Ascomycetes
6. yeasts
7. trichogyne
8. asci
9. basidiospores
10. sexual
11. lichen
12. mushroom
13. clamp connections
14. gills
15. ergot
16. zygomycetes
17. meio-blastospores
18. animals
19. slime molds
20. glucan

Putting Your Knowledge to Work

1. B. an increase in the rate of die off of forest trees
2. D. a plasmodial slime mold
3. C. The hyphae had clamp connections.

Doing Botany Yourself

Characters that would be important in identifying mushrooms include:

color of the cap
color of the gills
color and shape of spores
presence of any unique morphological features like a cup, a veil, or annulus.

To determine the color and morphology of the spores you will need to make a spore print. To do this remove the cap from a mushroom to be identified. Place it gill side down on a white piece of paper. Cover it with a glass jar and let it rest over night. Remove the jar and the cap. On the paper will be a pattern left by the spores. The color of the pattern is the color of the spores. Individual spores can be mounted and examined.

CHAPTER 27 ALGAE

After You Have Read the Chapter

1. What types of characters are used to classify algae?
2. At what point in the life cycle of algae can meiosis occur?
3. Compare the types of asexual reproduction found in algae.
4. What features are shared by some green algae and plants?
5. Discuss some of the features of kelp that resemble the leaves, stems, and roots of vascular plants.
6. What features of the red algae make them unique among the algae?
7. What kinds of spores do algae produce?
8. Why are the euglenoids, dinoflagellates, and cryptophytes studied by botanists and zoologists?
9. What are the possible sources for chloroplasts in the various groups of algae?
10. Describe the ecological roles of algae.
11. How are algae used commercially?

Constructing a Concept Map

Using the explanation in the introduction to the Study Guide, construct a concept map of the main ideas in the chapter. Use the headings in the text as source material.

Fill-in-the-Blanks

1. Algal spores that lack flagella are called ____________. (p. 636)
2. In the life cycles of many algae the ____________ is the only diploid phase. (p. 636)
3. Gametes of the different sizes, shapes, and/or different numbers of flagella are________________.
4. A(an) ____________ life cycle would include the fusion of a motile and a nonmotile gamete. (p. 637)
5. Cells of unicellular green algae often have a __________________ with a pyrenoid surrounded by starch granules and a _____________________ for light reception. (p. 638)
6. The shared biochemical pathway of flavonoid biosynthesis positions the _____________________ as the closest relatives of green plants. (p. 639)
7. In the genus *Chlamydomonas* the vegetative cells can function as ____________ during sexual reproduction. (p. 638)
8. *Volvox* can form small ____________ colonies inside the parent colony. (p. 640)
9. When mitosis is not followed by cytokinesis as in *Hydrodictyon,* ____________ cells can result. (p. 641)
10. In *Oedogonium* the egg forms in an enlarged cell called the ____________. (p. 642)
11. *Cladophora* is a(an) ____________ species because the sporophyte and the gametophyte look the same. (p. 642)
12. Stoneworts have ____________ antheridia and oogonia as do green plants. (p. 643)
13. *Ulva* has three types of _____________________—a sporophyte,female gametophyte, and a male gametophyte. (p. 644)
14. Current taxonomic analysis of the Charaphycean algae and green plants suggest they should be placed into __________________ taxonomic group(s). (p. 645)
15. *Ectocarpus* gametophytes produce ____________ gametangia on lateral branches. (p. 646)
16. Some brown algal gametes secrete ____________ to attract motile gametes. (p. 648)
17. The largest algae are ____________ which form coastal forests or large floating masses on the ocean. (p. 648)
18. A phycobilin pigment called ____________ gives red algae their red color. (p. 649)
19. In the red algal life cycle nonmotile ____________ fuse with the trichogyne of the carpogonium. (p. 650)

20. The carposporophyte of red algae produces ______________________ by mitosis that develop into the ______________________. (p. 650)
21. Diatoms have two a cell wall formed from two _________________ of glass bearing intricate patterns. (p. 652)
22. Diatoms store carbohydrate as ______________________ which is similar to the laminarin of the brown algae. (p. 652)
23. Euglenoids and cyptomonads have a ____________ instead of a cell wall. (p. 654)
24. ______________________ have two flagella and nuclei that contain chromosomes lacking histones and unusual mitosis. (p. 654)
25. Algae are the primary ______________________ in water ecosystems. (p. 657)
26. Algae in the oceans provide food and ____________ to other organisms. (p. 657)
27. ____________ are blooms of dinoflagellates that secrete toxins that are harmful to vertebrates. (p. 658)
28. ____________ formed from marine deposits of diatoms is used in abrasives and pool filters. (p. 658)
29. ____________ is an algal polysaccharide used as a stabilizer and emulsifier in creamy foods. (p. 659)
30. ___________________, a red alga, is used as a sushi wrapping and to add flavor in soups and salads. (p. 659)

Putting Your Knowledge to Work

1. *Chara* does not share which of the following with green plants?
   A. a body with node-internode construction
   B. leaf-like organs with a blade/petiole construction
   C. phragmoplastic cytokinesis
   D. multicellular sex organs

2. A phycologist retrieves a sample of *Ectocarpus* from a dive. She examines the specimen and finds unilocular sporangia on the tips of lateral branches. What will emerge from these structures?
   A. zoospores
   B. haploid eggs
   C. immature gametophytes
   D. isogametes

3. An unmanned submerged vehicle descends a cliff off the Atlantic Coast. The last algae it will find on the cliff as it descends will probably be
   A. Chlorophyta.
   B. diatoms.
   C. Phaeophyta.
   D. Rhodophyta.

4. A cytological and chemical analysis of a single-celled alga shows the presence of cellulose, chlorophyll c, and fucoxanthin. This alga is most likely a
   A. dinoflagellate.
   B. brown alga.
   C. cryptomonad.
   D. diatom.

Doing Botany Yourself

Develop an experimental protocol to identify the relative numbers of algal cells in samples of marine plankton.

Answer Key to Chapter 26 Study Guide Questions

After You Have Read the Chapter

1. Habitat, pigmentation, carbohydrate storage, cell wall components, and flagella are used to classify algae. (p. 635)
2. Meiosis may be sporic, gametic, or zygotic. (p. 637))
3. Asexual reproduction may include fragmentation, vegetative propagation, or mitotic spores. (p. 637-638)
4. Some green algae and plants both retain the zygote in the gametophyte, similar flavonoid biosynthesis, similar ribosomal RNA sequences, phragmoplastic cell division, unique flagellar ultrastructure, similar chloroplasts, and photorespiration. (p. 638-644)

5. The blades resemble leaves, the stipe with its sieve cells resembles a stem with vascular tissue, and the anchoring holdfast resembles a root system. (p. 648)
6. The red algae have chloroplasts with a single thylakoid membrane per band and phycobilins like the cyanobacteria. They have no motile cells and produce nonmotile male gametes, spermatia, and eggs. Multicellular forms have a complex three-stage life cycle including gametophyte, carposporophyte, and tetrasporopyte. (p. 649-650)
8. Algae can make zoospores, aplanospores, carpospores, tetraspores, or auxospores depending on the species. (p. 636, 637, 642, 645, 648, 650, 652)
9. In the green algae and euglenoids the chloroplasts seems too have a Chloroxybacteria origin from an alga with chlorophyll a and chlorophyll b. Red algal chloroplasts and chloroplasts of cryptomonads may have had a cyanobacterial ancestor with chlorophyll a and phycobilins. Heliobacterium seems to be a possible ancestor for chloroplasts of the brown algae, Chrysophyta, and the dinoflagellates with chlorophyll a and c. (p. 656-657)
10. The most important ecological role of algae is that of plankton as the producers and in aquatic ecosystems and as generators of 50-70% of atmospheric oxygen. Red and brown algae form forests in intertidal zone providing food and protection for many animals. coralline algae form coral reefs. Under conditions of pollution algae can form blooms that threaten water supplies and kill animals. Red tides formed by dinoflagellates can poison shellfish supplies and cause massive fish kills. (p. 657-658)
11. Algae are used as sources of food, stabilizers, emulsifiers, bacterial media, abrasives, and filtering agents. (p. 659)

Constructing a Concept Map

Compare your concept map to the sample in the introduction to the Study Guide.

Fill-in-the-Blanks

1. aplanospores
2. zygote
3. anisogametes
4. oogamous
5. chloroplast, stigma
6. Charaphyceae
7. isogametes
8. daughter
9. coenocytic
10. oogonium
11. isomorphic
12. multicellular
13. adult forms
14. one
15. plurilocular
16. ectocarpene
17. kelps
18. phycoerythin
19. spermatia
20. carpospores, tetrsporophyte
21. valves
22. chrysolaminarin
23. periplast
24. Dinoflagellates
25. producers
26. oxygen
27. Red tides
28. Diatomaceous earth
29. Carrageenan
30. nori

Putting Your Knowledge to Work

1. B. leaf-like organs with blade/petiole construction
2. A. zoospores
3. D. Rhodophyta.
4. D. diatom.

Doing Botany Yourself

The major groups of marine plankton are the green algae, diatoms and other chrysophytes, and dinoflagellates. All eukaryotic photoautotrophs use chlorophyll *a* as their photosynthetic pigment. Chlorophyll *a* absorbs light maximally at a unique range of wavelengths. This absorption can be used to identify the relative numbers of algal cells in plankton samples.

Chlorophyll *a* absorbs maximally at 432 nm and 662 nm. Samples can be placed in a spectrophotometer and illuminated with one of these wavelengths. Samples with more cells will have a higher absorbance than samples with fewer cells. This procedure will allow you to rank samples from fewest to most cells, but not actually to count the number of cells present.

CHAPTER 28 BRYOPHYTES

After You Have Read the Chapter

1. Describe the survival challenges faced by the earliest land plants and the adaptations that evolved to meet them..
2. Describe in words a typical bryophyte life cycle.
3. Distinguish the members of the three bryophyte groups using morphology of the gametophytes.
4. Describe the types of sporophytes produced by bryophytes.
5. What types of spore dispersal mechanisms are utilized by bryophytes?
6. What is the role of the protonema in the bryophyte life cycle?
7. What are rhizoids and what is their function?
8. Compare leptoids and hydroids to vascular plant xylem and phloem.
9. Where can bryophytes be found in nature?
10. What factors can influence the development of bryophytes?

Constructing a Concept Map

Using the explanation in the introduction to the Study Guide, construct a concept map of the main ideas in the chapter. Use the headings in the text as source material.

Fill-in-the-Blanks

1. Unlike vascular plants, bryophytes lack ________________________ tissue or lignified cells. (p. 664)
2. Bryophytes are attached to the substrate by hairlike ____________________. (p. 664)
3. Bryophytes produce ____________________ in antheridia formed on gametophytic thalli. (p. 665)
4. Each bryophyte archegonium produces one ________________________. (p. 665)
5. Bryophytes require ____________________ for fertilization because they have swimming sperm. (p. 666)
6. __________________ have leafy, radially symmetrical gametophytes whose leaves have a midrib. (p. 667)
7. Absorptive, sterile filaments called __________________ intermix with moss archegonia and antheridia. (p. 669)
8. The operculum of the moss __________________ falls off when it is mature allowing the spores inside to be released. (p. 669)
9. The moss spore germinates into a filamentous ____________________ from which the mature gametophyte will develop. (p. 669)
10. A moss archegonium consists of a long, slender ________________ and a swollen base or venter that encloses the egg . (p. 670)
11. The developing moss sporophyte is protected by the ________________________ that develops from the expanding venter. (p. 670-671)
12. The moss ________________ consists of a foot, , seta, and a ____________. (p. 671)
13. The moss sporangium has a central region of sterile tissue, the ______________________, and releases its spores through a toothed peristome. (p. 671)
14. Liverworts were once thought to be helpful in treating ____________________ailments. (p. 672)
15. *Marchantia* ____________________are umbrella-shaped stalks that grow up from the upper surface of the thallus. (p. 674)

16. Liverworts reproduce asexually by fragmentation and by the production of _________________ in cups on the upper surfaces of the gametophytes. (p. 674)
17. The sporangia of liverworts contain spores and sterile _________________ that assist in the dispersal of the spores. (p. 675)
18. The thalli of hornworts have mucilage-filled cavities that contain nitrogen-fixing bacteria and one ____________ per thallus cell with an algal-like pyrenoid. (p. 676)
19. The water-holding capacity of peat mosses is due to rows of dead ______________ in their leaves. (p. 678)
20 Bryophytes first appear in the fossil record in the ____________. (p. 681)
21. One commercial use of bryophytes is as a renewable source of fuel in the form of ________________. (p. 682)
22. Bryophyte gametophyte development can be influenced by __________________ quality and day length. (p. 680)
23. The plant hormone _______________________ can regulate protonema and gametophyte development. (p. 680)
24. Bryophytes and ___________________________ plants are thought to have diverged from a common ancestor more than 430 million years ago. (p. 680)

Putting Your Knowledge to Work

1. A radiation biologist is studying the release of radioactivity from a nuclear power plant in Alaska. Which of the following organisms would make a good choice to monitor the release over several years?
   A. annual herbs
   B. Soil fungi
   C. soil bacteria
   D. Mosses

2. A botanist removes a capsule from a moss sporophyte. In tissue culture the tissue differentiates into a bisexual gametophyte. This species normally has unisexual gametophytes. What is the best explanation of the results?
   A. Bisexuality in this moss is a recessive character.
   B. Sexuality in this moss is regulated by two dominant alleles.
   C. Sexuality in this moss is regulated by sex chromosomes.
   D. The capsule cells underwent a series of sequential mutations to a gametophyte.

3. Which of the following characters would you expect to find in bryophytes and vascular plants?
   A. stomates on the gametophyte
   B. a protonema
   C. a gametophyte retained in the sporophyte
   D. a sterile layer around the fertile cells of the sporangium

4. Which of the following is the best evidence that bryophytes are not the ancestors of vascular plants?
   A. Bryophytes lack flavonoid biosynthesis.
   B. The retention of the zygote in the body of the sporophyte in both bryophytes and vascular plants.
   C. The earliest fossil bryophytes date only to the Carboniferous.
   D. Bryophyte sporophytes lack stomates.

5. What is the possible ecological significance of the requirement of red light for the production of buds on moss protonema?
   A. to assure germination of spores on or close to the surface, not deep in the soil
   B. to assure that moss spores germinate only on exposed rocks
   C. to assure that spores germinate only where they occur uncovered by piles of other spores
   D. to prevent germination of spores in the winter when light levels are low

Doing Botany Yourself

Some botanists theorize that mechanical pressure is required, in part, in order for a moss zygote to develop into a sporophyte. Design an experiment to test this hypothesis.

Answer Key to Chapter 28 Study Guide Questions

After You Have Read the Chapter

1. Challenges to early land plant survival included keeping wet and reproduction using swimming sperm. Some plants, vascular plants, developed a thick cuticle and vascular tissue to get, transport, and retain water. Bryophytes developed strategies that incorporate slow growth and small size to decrease water demand. Land plants also developed ways to release gametes only when water was available for sperm transport. (p. 664)
2. A haploid spore germinates into a leafy or thalloid gametophyte. The gametophyte produces antheridia-bearing sperm and archegonia, each with one egg, on one or more separate thalli. Sperm produced by mitosis are released to swim to the egg, which is also produced by mitosis, in the archegonium. Fertilization occurs in the archegonium and the resulting zygote divides to form the new sporophyte. The sporophyte is retained in the gametophyte and depending on the type of bryophyte can be wholly or partly parasitic. The sporophyte capsule produces spores by meiosis. Each spore when released can produce new gametophytes either through a protonema or directly. (p. 666)
3. Mosses have radially-symmetrical leafy gametophytes produced from filamentous protonemas that germinate from spores. Stalked archegonia and antheridia may be produced on the same or separate plants. Liverworts may have leafy gametophytes with three-ranked leaves or thalloid gametophytes. Archegonia and archegonia may be sunken in the gametophyte tissue or borne on specialized gametophores. Hornworts have thallose gametophytes whose cells have one chloroplast each and have mucilage-filled cavities. The archegonia are not distinct from the thalloid tissue. (p. 669, 674, 676)
4. Moss sporophytes have a foot, a long seta, and a capsule. In most mosses the capsule has an internal portion of sterile tissue called the columella and outer layers of spore-producing cells. The tip of the capsule is covered by an operculum that covers the peristome. The teeth of the peristome and contraction of the capsule assist in spore dispersal. The sporophytes are dependent on the gametophyte for nutrition. Liverwort sporophytes lack a seta, are mostly dependent on the gametophyte for nutrition, and lack a peristome. The capsule has no columella, and elators among the spores aid in spore dispersal. Hornwort sporophytes have a basal meristem, produce spores for many months, are almost nutritionally independent of the gametophyte, have stomates on the capsule, lack a seta, and release spores by disintegration of the tip of the capsule. (p. 67-671, 675,677)
5. In mosses spore dispersal is aided by pressure that develops in the capsule as the capsule dries. The pressure can force the operculum off the capsule. In species with a peristome the spores are then released through pores in the peristome or flung out by the wetting and drying of the peristome teeth. In peat mosses the "explosion" of the capsule blows off the operculum and forcibly expels the spores. In liverworts hygroscopic elators produce violent twisting forces when they dry forcing spores from the capsule. In hornworts spores are released passively as the tip of the sporangium decay away. (p. 671, 675, 677)
6. The protonema is produced from the germinating spore. It allows for the production of several gametophytes from a single spore, increasing the chances of the survival of the gametophyte. (p. 669)
7. Rhizoids are unicellular or multicellular outgrowths of the lower surface of the gametophyte. They anchor the plant and in some cases aid in absorption of water and minerals. (p. 664)
8. They are found in the stems of large mosses. Leptoids transmit nutrients much as phloem does, and hydroids transport water much as xylem does. (p. 670)

9. Bryophytes of some type can be found nearly everywhere on the planet on bare soil, rocks, the forest floor, and branches, from the Antarctic to desert regions. They are widespread, but each species requires a rather specific habitat. (p. 678-679)
10. Bryophyte development can be influenced by light, hormones, or other organisms such as bacteria and cyanobacteria. (p. 680-681)

Constructing a Concept Map

Compare your concept map to the sample in the introduction to the Study Guide.

Fill-in-the-Blanks

1. vascular tissue
2. rhizoids
3. Sperm
4. egg
5. Water
6. mosses
7. paraphyses
8. sporangium
9. protonema
10. neck
11. calyptra
12. sporophyte, capsule
13. Columella
14. liverworts
15. archegoniophores
16. gemmae
17. Elators
18. Chloroplast
19. cells
20. Carboniferous
21. peat
22. light
23. auxin
24. vascular plants

Putting Your Knowledge to Work

1. D. mosses
2. C. Sexuality in this moss is regulated by sex chromosomes.
3. D. A sterile layer around the fertile cells of the sporangium
4. C. The earliest fossil bryophytes date only to the Carboniferous.
5. A. to assure germination of spores on or close to the surface, not deep in the soil

Doing Botany Yourself

The retention of the zygote in the gametophyte provides mechanical pressure on the developing zygote. If fertilization were done with free sperm and eggs surgically removed from the archegonium, then the factor of mechanical pressure would be removed. If free-formed zygotes developed into diploid gametophytes and not into sporophytes, this would be an indication that the differences in form were related to mechanical pressure.

Further experimentation would involve fertilizing eggs in a very firm agar medium that would mimic the gametophyte tissue. The firmness of the medium could be changed and the influence of these changes on the developing sporophytes observed. If a firm medium produced more normal sporophytes and a looser medium sporophytes that were more gametophyte-like. this would be further evidence that pressure is a factor in sporophyte development.

CHAPTER 29 SEEDLESS VASCULAR PLANTS

After You Have Read the Chapter

1. Using morphological and reproductive features define seedless vascular plants as a distinct group.
2. Describe a generalized seedless vascular plant life cycle.
3. Describe the various arrangements of vascular tissue found in seedless vascular plants.
4. Compare microphylls to megaphylls.
5. What features of the wisk ferns make them difficult to place in the plant kingdom?
6. Contrast *Lycopodium* and *Selaginella.*
7. What features of horsetails are unique among living vascular plants?
8. Describe the features of fern leaves that distinguish them from the leaves of other seedless vascular plants.
9. What are the current theories about the origins of the seedless vascular plant groups?
10. What types of external and internal factors affect the development of seedless vascular plants?

Constructing a Concept Map

Using the explanation in the introduction to the Study Guide, construct a concept map of the main ideas in the chapter. Use the headings in the text as source material.

Fill-in-the-Blanks

1. Seedless vascular plants have xylem and other tissues with cell walls impregnated with _________________that provides support to the plant body. (p. 688)
2. The ____________________ in the seedless vascular plant life cycles is nutritionally independent from the gametophyte. (p. 689)
3. In seedless vascular plants the ______________________ is eithers mall, thin, green or subterranean with a symbiotic fungus. (p. 689)
4. Many seedless vascular plant produce _________________________- consisting of collections of sporophylls at the tips of stems. (p. 690)
5. Seedless vascular plant gametophytes are usually bisexual bearing archegonia that contain eggs and ______________ that produce sperm. (p. 689)
6. A pith is lacking in stems with a __________________________. (p. 691)
7. At the nodes, horsetails have a ___________with a ring of vacular bundles around a pith at the nodes.(p. 691)
8. Megaphylls have many branched veins and were derived from ______________________. (p. 684)
9. The stems of *Psilotum* are photosynthetic and lack true ______________________. (p. 693)
10. *Selaginella* is considered to be _____________because it produces both megaspores and microspores. (p. 695)
11. The club mosses bear sporangia on leaves called ____________________________. (p. 695)
12. Quillworts are unusual lycopods because have _______________ photosynthesis and secondary vasuclar tissue. (p. 696)
13. The unusual features of ____________ include vestigial leaves, jointed stems, and cell walls embedded with glass. (p. 698-699)
14. Horsetail spores have four spoon-shaped _________________ that help in spore dispersal. (p. 699)
15. The __________________ may play a role in determining the development of the gametophytes of horsetails into monecious or dioecious plants . (p. 699)
16. The _________ of the fern sporangium aids in the forcible ejection of spores from the sporangium. (p. 702)
17. The form of the fern body is defined by its ___________________that develop from a rhizome throughout the growing season. (p. 700)

18. Homosporous fern gametophytes bear antheridia and archegonia on the ________________ surface. (p. 702
19. Young fern leaves emerge in the form of ____________________ that unfold by circinate vernation into mature leaves. (p. 700)
20. Heterosporous ferns like *Azolla* bear their sporangia in hard cases called ______________________. (p. 704)
21. Spore germination in ferns can be initiated by the interaction of phytochrome and ___________. (p. 704)
22. ___________ secreted by gametophytes regulate the development of antheridia in other nearby gametophytes. (p. 704-705)
23. The sporangia of the Zosterophyllophyta were borne ___________________on the stem at the tips of branches. (p. 701)
24. Rhyniophytes and Trimerophytes had ________________________ sporangia. (p. 711)
25. The ___________________ were the probable ancestors of all vascular plants except the lycopods. (p. 711)

Putting Your Knowledge to Work

1. A botanist finds a small specimen while examining surface soil samples in the Amazonian rain forest. The samples resemble pieces of rhizomes but they lack roots. Closer examination reveals an endophytic fungus and archegonia and antheridia. This specimen is most likely the gametophyte of a
   A. homosporous lycopod.
   B. horsetail.
   C. quillwort.
   D. heterosporous fern.

2. A new plant has been discovered. It has small leaves that alternate with buds at the nodes. Sporangia are borne in terminal strobili on green stems. This plant is most likely a relative of the
   A. ferns.
   B. horsetails.
   C. psilotophytes.
   D. lycopods.

3. You need to culture as many horsetail sporophytes as possible from a sample of spores. To do this you would use culture conditions that were
   A. cool with intense red light.
   B. warm and dark.
   C. warm with soft blue light.
   D. cool with intense blue light.

4. Cladistic analysis places the psilotophytes nearest the vascular plant ancestor on the basis of a lack of roots. Molecular genetic analysis groups together the bryophytes (also lacking roots) and the lycopods and separates both from all other vascular plants. How could this seeming conflict be explained?
   A. The lack of roots is always a primitive condition.
   B. The lack of roots is a shared-derived condition in all vascular plants.
   C. The psilotophytes, lycopods, and bryophytes share one ancestor.
   D. The lack of roots may be a secondary adaptation of a rooted ancestor.

5. The ancestor of all vascular plants probably had
   A. megaphylls.
   B. sporangia arranged in strobili.
   C. homosporous sporangia.
   D. vascular tissue in a siphonostele.

Doing Botany Yourself

Devise an experiment using horsetail gametophytes to test what per cent of the zygotes are the result of self-fertilizations.

Answer Key to Chapter 28 Study Guide Questions

After You Have Read the Chapter

1. All have the features of green plants including an alternation-of-generations life cycle and similar photosynthetic pigments, store food as starch, phragmoplastic cell division, cellulose cell walls,and sterile layer around fertile cells in sporangium. In addition, they have a well-developed cuticle, vascular tissue lignified secondary cell walls, nutritionally independent sporophytes and gametophytes, long-lived, dominant sporophytes, and motile sperm;. (p. 688)
2. All seedless vascular plants have an alternation of generations type life cycle with a dominant soprophyte and free-living gametophyte. The sporophyte produces sporangia that generate either one type of spores or two types of spores(microspores or megaspores)by meiosis that develop into either unisexual or bisexual gametophytes bearing the appropriate sex organs, archegonia and/or antheridia. Some species of seedless vascular plants produce subterranean gametophytes that lack chlorophyll and depend on an endophytic fungus to survive. Other species produce thin, green, photosynthetic gametophytes. Sperm produced in the antheridia swim to the egg produced in each archegomium. The resulting zygote begins development in the tissues of the gametophyte, but the sporophyte quickly becomes an independent plant. (p. 689-690)
3. Vascular plants may have protosteles (a solid core of xylem surrounded by a ring of phloem), siphonsteles(a parenchyma core or pith surrounded by a inner ring of xylem and an outer ring of phloem, dictyosteles (a siphonostele with a high number of leaf gaps), or eusteles(a ring of vascular bundles of xylem and phloem around a pith). (p. 691)
4. Microphylls have a single vein that stops at the base and probably evolved from enations. They vary in size from very small to more than a meter in length. Megaphylls have extensively branched venation systems and probably were derived from branching systems. They vary in size from microscopic to more than a meter long. (p. 691)
5. Wisk ferns have a complex of features unique among vascular plants including a lack of roots(they hava a rhizome with rhizoids), prophylls instead of leaves, and photosynthetic stems, and 3-lobed synangia. They alsoproduce a distinct class of flavenoids and contain the same chlroroplast genome type as horsetails, ferns, and seed plants. (p. 693)
6. *Lycopodium* bears homosporous sporangia on the surfaces of sporophylls. The sporophylls may have the same structure as vegetative leaves or specialized and collected into strobili. The spores germinate into subterranean, free-living gametophytes bearing antheridia and archegonia. *Selaginella* bears microsporangia and megasporangia on sporophylls collected into strobili. The megasporangia bear four megaspores and the microsporangia hundreds of spores each. The germinated megaspore develops into a megagametophyte-bearing archegonia within the spore wall. The microspore cell divides to form a single outer sterile layer of cells and fertile internal cells. Each fertile cell becomes a motile sperm. (p. 695)
7. Horsetails have jointed, grooved stems with a strong node and internodal arrangement. The stems have vestigial leaves that alternate with branches or buds. The stem cells are impregnated with glass. The stems have a unique internal canal system. The sporangia are attached to unique sporangiophores and borne in cones. The spores have spoon-shaped elators and germinate into gametophytes with variable sex. (p. 697-699)
8. The leaves of ferns are the major organ of the plant body. The stem or rhizome is reduced or underground. The leaves display circinate vernation as they develop. The sporangia are borne on the undersides of the all the leaves in masses called sori. An indusium or covering for the sorus grows out of the leaf surfface in many fern species. (p. 700)
9. The first fossil vascular plants appeared in the Silurian and resembled *Cooksonia* with a dichotomaously-branched axis and little differentiation between plant vegetative organs. The Zosterophyllophytes with their lateral sprorangia, exarch xylem are thought to be the ancestors of the lycopods. The Rhiniophytes, including *Cooksonia* represent another evolutionary line of plants with terminal sporangia and endarch xylem. the Trimerophytes, the probable ancestors of the horsetails, ferns, and Psilotophytes, evolved from the Rhiniophytes. (p. 709-712)

10. Seedless vascular plant organs, gametophytes, and spores can respond to red light, blue light, temperature, hormones, and fungal partners. (p. 705-706)

Constructing a Concept Map

Compare your concept map to the sample in the introduction to the Study Guide.

Fill-in-the-Blanks

1. lignin
2. sporophyte
3. gametophyte
4. strobili
5. antheridia
6. protostele
7. eustele
8. branching systems
9. leaves
10. heterosporous
11. sporophytes
12. CAM
13. horsetails
14. elators
15. environment
16. annulus
17. leaves
18. lower
19. fiddleheads
20. sporocarps
21. red light
22. Antheridogens
23. laterally
24. terminal
25. Trimerophytophyta

Putting Your Knowledge to Work

1. A., homosporous lycopod
2. B., horsetails
3. A., cool with intense red light
4. D.,The lack of roots may be a secondary adaptation of a rooted ancestor.
5. C., homosoporous sporangia

Doing Botany Yourself

First it would be necessary to determine that the gametophytes could be self-fertile and the percent of archegonia that produce viable zygotes. To do this one would need to germinate spores individually on agar in sterile culture and grow each gametophyte under environmental conditions that permote bisexual gametophytes. The sexually mature gametophytes would be observed for the number of archegoina produced and the number of zygotes that form. The zygotes could be removed and cultured to maturity. The ratio of zygotes formed to archegonia produced would be the frequency of self-fertilization under non-competing conditions from other sperm. Gametophytes from different genetic stock would need to be placed in the same sealed culture. The number of archegonia counted, and the number of zygotes produced per gametophyte. The zygotes would need to be cultured to form gametophytes. If more zygotes are formed then by self-fertilization alone, then cross-fertilization is probably occurring. The difference between the number of zygotes with only self-fertilization and cross-fertilization is a rough estimate of the frequency of cross-fertilization. Genetic and molecular studies could be performed to determine more accurately the portion of the gametophytes that were produced from cross-fertilization and what percentage from fertilization.

CHAPTER 30 GYMNOSPERMS

After You Have Read the Chapter

1. What characters are shared by all gymnosperms?
2. Describe the types of cells that can be present in a gymnosperm pollen grain?
3. Describe the distinctive features of the gymnosperm seed.
4. Describe the development of the gymnosperm megasporangium in the ovule.
5. In what ways are seeds borne by different species of gymnosperms?
6. Describe the different pollen organs of gymnosperms.
7. What is a pollination droplet and what is its function?
8. Compare the Ginkgophyta, Cycadophyta, Pinophyta, and Gnetophyta on the basis of leaves, seeds, and pollen grains.
9. How are gymnosperms used commercially?
10. What features are shared by the Progymnosperms and Coniferophytes?

Constructing a Concept Map

Using the explanation in the introduction to the Study Guide, construct a concept map of the main ideas in the chapter. Use the headings in the text as source material.

Fill-in-the-Blanks

1. Microspores that germinate from the side opposite the tetrad scar are considered to be ________________. (p. 718-720)
2. The pollen grain contains the ______________________. (p. 718)
3. In gymnosperms the ______________________ are carried to the ovule by the pollen tube. (p. 718)
4. Conifers produce pollen cones and ______________________ cones. (p. 720)
5. The gymnosperm female gametophyte develops in the ________________________. (p. 721)
6. In most gymnosperms embryonic development begins with a period of ___________ ______________ division. (p. 721)
7. One zygote can form more than one embryo by the process of ________________ polyembryony. (p. 721)
8. Many gymnsoperms produce ____________________seed cones that evolved from seed-bearing branch systems. (p. 721)
9. The pollen tubes of *Ginkgo* and other gymnosperms are ________________ meaning that they can obtain nutrition by digesting and absorbing the tissues of the female gametophyte. (p. 727)
10. Some cycad _____________ grow near the surface of the soil and contain nitrogen-fixing cyanobacteria. (p. 727)
11. Cycads, like *Ginkgo,* have ________________ sperm even though they also produce pollen tubes. (p. 727)
12. Cycads have large, palm-like __________________ produced in a crown near the top of the stem.(p. 727)
13. Pines bear needle-like __________________in clusters called fascicles.(p. 728)
14. From ________________to the production of mature seeds in pines can take from 14 to 20 months.(p. 730)
15. The ________________ is the geographic region with the fewest species of gymnosperms. (p. 731)
16. Gnetophytes share many features with ________________ including vessels and double fertilization. (p. 730)
17. Wood from white spruce produces paper pulp and from Douglas fir produces______________________. (p. 731)

18. The resin produced in the resin canals of conifers consists of liquid ______________________ and rosin, a waxy solid .(p. 731)
19. Fossilized dammar resin forms ______________________, the only jewel produced by a plant.(p. 732)
20. The most likely ancestor of the gymnosperms is the __________________________of the Deveonian and Carboniferous periods. (p. 733)
21. Seed ferns are Devonian plants that produced fern foliage that bore ________________________. (p. 734)
22. The sporophyte of ______________________ resembled those of cycads, but the cones were bisexual. (p. 735)
23. The _____________________ were Carboniferous Pinophyta with strap-like leaves and compound strobili. (p. 736)
24. The extinct _____________________ had a combination of Araucarian and pine-like features. (p. 736)
25. Cladistic and gene sequencing support a shared ancestor between the gnetophytes and ________________. (p. 737)

Putting Your Knowledge to Work

1. Gnetophytes and angiosperms are the only two groups that have vessels in the xylem and double fertilization. Some botanists use these features to point to a common ancestor for the two groups. What is another possible explanation for these similar, unique features in the two groups besides common ancestry?
   A. hybridization between gnetophytes and angiosperms
   B. adaptation to similar environmental conditions
   C. gnetophytes and angiosperms each crossing with the same extinct species
   D. special creation of each group at different times to fill similar niches

2. The pollen/seed mechanisms for reproduction are an adaptive advantage for land plants by
   A. keeping offspring near parents where conditions favor survival.
   B. eliminating the need for water for reproduction
   C. allowing for asexual reproduction of diploids
   D. assuring that male and female gametophytes are produced on the same plant.

3. The long reproductive cycle of many conifers excludes them from the floras of
   A. relatively stable habitats.
   B. tropical regions
   C. the Great Plains.
   D. highly disturbed sites.

4. A female pine cone formed in April 1996 will be able to release mature seeds in
   A. August 1996.
   B. April 1996.
   C. September 1997.
   D. January 1997.

Doing Botany Yourself

Develop a dichotomous key to the conifers in your area.

Answer Key to Chapter 30 Study Guide Questions

After You Have Read the Chapter

1. Gymnosperms are seed plants in which the pollen is delivered directly to the ovule, they have no stigma, and the seeds are not enclosed in a fruit. Gymnosperms have secondary xylem usually composed of only tracheids, but some gnetophytes have vessels. Gymnosperms are mostly woody shrubs or trees, but a few are woody vines. All other characters of gymnosperms vary from group to group. (p. 718)
2. A gymnosperm pollen grain may have one or two, or no, prothallial cells. It has a generative cell that divides to form the stalk cell and a body cell that divides to form the two sperm cells. (p. 718-720)
3. A gymnosperm seed consists of a single outer covering or integument, a female gametophyte that comprises most of the seed, and one or more embryos. (p. 720)
4. A single cell in the megasporangium of the ovule undergoes meiosis to form a linear tetrad of megspores. Three disintegrate. In all but Gnetum, the nucleus of the remaining one, the functional megaspore, undergoes many mitotic divisions to form a free-nuclear female gametophyte with from 256-8,00 nuclei depending on the species. Cell walls form around each nucleus prior to the formation of archegonia. From 2-200 archegonia form at the micropylar end of the gametophyte. Each forms a single egg. One or more eggs can be fertilized. In some gnetophytes archegonia do not form. (p. 721)
5. Gymnosperm seeds can be borne singly on the terminal end of a stalk. In ginkgos this seed is surrounded by a fleshy integument. In the yew family a red, fleshy aril surrounds the seed. Seeds can also be borne in strobili that vary in size from less than one centimeter across, as in junipers, to up to one meter long, as in some cycads and pines. The seeds in cycad strobili are borne on modified leaves and in pines on modified shoots. Gnetophytes bear seeds in reduced, complex cones. (p. 721)
6. *Gingko* pollen cones are formed from spirally arranged stalks bearing paired pollen sacs. Pines bear microsporangia on the abaxial surface of microsporophylls collected into cones. Cycads have microsporangia on the adaxial surface of sporophylls collected into cones. Gnetophytes have compound microstrobili with pollen sacs borne terminally on shartstalks subtended by bracts. (p. 713–14)
7. Gymnosperm ovules exude a sticky droplet from the micropyle at the time pollen is being shed. Pollen, usually carried to the ovule by the wind, sticks to the droplet. As the droplet dies it pulls numerous pollen grains into the pollen chamber and into direct contact with the female gametophyte. (p. 726)
8. Gingkophyta: leaves are thin blades with dichotomous venation, seeds have a fleshy outer layer and a hard inner layer, and pollen grains have a haustorial pollen tube and motile sperm. Cycadophyta: seeds have a fleshy outer layer with a papery inner integument layer, leaves are large and compound, resembling the leaves of palm trees; the pollen tube is haustorial, and motile sperm are released from the pollen grain, not the pollen tube. Pinophyta: leaves are simple, needle-like, awl-like, or scalelike borne in fasicles in pine and singly in other groups, seeds are borne in cones or singly. Cones may be woody as in pine or composed of a few fleshy scales as in junipers; seeds may or may not have a wing as in pines; pollen is haustorial but transports the sperm, which are nonmotile, to the ovule. The pollen may have bladders as in pine. Gnetophyta: leaves vary from angiosperm-like in *Gnetum* to very reduced and vestigial in *Ephedra* to only two per plant with basal meristems as in *Welwitschia;* pollen has no prothallial cells and delivers the sperm to the ovule; double fertilization can occur. (p. 727-731)
9. Gymnosperms are used to produce lumber and in the manufacture of paper, furniture, rosin, turpentine, deodorants, shaving lotions, drugs, lemon flavor, varnishes, and linoleum. (p. 731-732)
10. Progymnosperms and gymnosperms both have heterosporous members, simple leaves derived from branching systems, seeds or seed-like structures, vascular cambium and large amounts of secondary xylem and phloem similar to modern gymnosperms. Xylem tracheids are long and have bordered pits. (p. 734)

Constructing a Concept Map

Compare your concept map to the sample in the introduction to the Study Guide.

Fill-in-the-Blanks

1. pollen grains
2. male gametophyte
3. sperm
4. seed
5. ovule
6. compound
7. cleavage
8. compound
9. haustorial
10. roots
11. flagellated
12. leaves
13. leaves
14. pollinatoin
15. seeds
16. angiosperms
17. lumber
18. turpentine
19. amber
20. progymnosperms
21. seeds
22. cycadeoids
23. cordaites
24. Voltziales
25. angiosperms

Putting Your Knowledge to Work

1. B. adaptation to similar environmental conditions
2. C. reproduction that bypasses fertilization
3. A. relatively stable habitats
4. C. September 1995

Doing Botany Yourself

Collect several samples of a variety of conifers that grow in your area. Examine the cones to identify characters that vary from species to species and characters that are unique to one or a few species. Use these characters to construct a dichotomous key.

CHAPTER 31 ANGIOSPERMS

After You Have Read the Chapter

1. What features of the Koonwarra angiosperm convinced paleobotanists it was not a fern?
2. Describe the current idea of the primitive angiosperm flower.
3. When and where are the angiosperms thought to have evolved?
4. What was the role of insects in the diversification of the angiosperms?
5. Describe the methods used to verify the classification of groups of angiosperms?
6. What is a flora and how is it used?
7. What is the most current theory as to the origin of flowers?
8. What is the most probable ancestral group for the angiosperms?
9. How is the chloroplast gene *rbc*L used to determine the phylogenetic relationships between plant groups?
10. What evidence supports a date for the divergence of monocots and dicots from a common ancestor over 200 million years ago?

Constructing a Concept Map

Using the explanation in the introduction to the Study Guide, construct a concept map of the main ideas in the chapter. Use the headings in the text as source material.

Fill-in-the-Blanks

1. The Koonwara angiosperm supports the theory that the ancestral angiosperms were not____________. (p. 742)
2. A diverse flora of angiosperms is found in the fossil record by the ________________________ period. (p. 745)
3. Some botanists theorize that the angiosperm ovule may have evolved from a seed fern that bore its seeds in a ____________. (p. 746)
4. The earliest flowers had gynoecia with ____________ carpels that developed into follicle or nutlet fruits. (p. 745)
5. Early angiosperm pollen was probably not distinguishable from that of ____________________. (p. 746)
6. The earliest angiosperms are thought to have evolved in the semiarid regions of central ________________________ leaving few fossils. (p. 747)
7. The carpel of angiosperms may have evolved from a folded ________________________. (p. 746)
8. Botanists now believe that pre-Cretaceous angiosperms were adapted to ____________, and ____________ climates. (p. 747)
9. By the early Tertiary period angiosperms were displacing _________________in lowland basin areas. (p. 747)
10. Early in seed-plant evolution ______________________ had become important pollinators. (p. 747)
11. Angiosperms evolved floral ____________________ and odors for attracting pollinators. (p. 747)
12. ___________________, possible close relatives of angiosperms, had bisporangiate cones and were specialized for beetle pollination. (p. 747)
13. Books that contain plant lists, keys, and descriptions are called ____________. (p. 748)
14. The ____________________ is the most important feature of angiosperms used in classification. (p. 749)
15. Analysis of the nucleotide sequences in ______________________ can provide vital information when evaluating the relationships between groups of plants. (p. 752)
16. According to DNA sequence analysis the ______________________ may have diverged from the dicots as long as 320 million years ago. (p. 753-754)
17. Cladistic analysis of the angiosperms recognizes ____________________ as the living flowering plant genus closest to the ancestral type. (p. 752)

18. According to the most recent theory of the origin of the flower, the flower evolved from a ___________________ that evolved from the fertile leaves of seed ferns. (p. 753)
19. Cladistic analysis combined with DNA sequencing, paleobotany, and classic morphological analysis suggests that angiosperms may share a common ancestor with the ___________ and _______________. (p. 753)
20. The ______________________ proposes that the amount of dissimilarity between the nucleic acid sequences of genes from different sources should be proportional to the length of time since each diverged from a common ancestor. (p. 753)

Putting Your Knowledge to Work

1. Which of the following supports the origin of angiosperms at least 200 million years ago?
   A. DNA sequence analysis of the chloroplast *rbc*L gene
   B. the 120-million-year age of the first angiosperm fossil flower
   C. shared features with cycadeoids
   D. frequent finds of dinosaur and angiosperm fossils in the same fossil beds in the Cretaceous

2. The direct angiosperm ancestor would not be expected to have:
   A. linear tetrad of megaspores.
   B. herbaceous growth habit.
   C. insect pollinators.
   D. compound carpels.

3. According to current theories about the origin of angiosperms, where should paleobotanists be looking for fossils of the angiosperm ancestor?
   A. regions of central US once shores of inland seas
   B. Andes Mountains
   C. far eastern regions of Canada
   D. the Amazon River Valley

4. If the angiosperms, cycadeoids, and gnetophytes are derived from a common ancestor, which of the following would the ancestor be likely to have?
   A. dioecious flowers
   B. bisporangiate stobili.
   C. carpels formed from folded leaves
   D. honeybee pollinators

Doing Botany Yourself

You are doing a cladistic analysis of fossil plants that could be ancestors of the angiosperms. Choose five characters that would be found in the ancestor and considered to be the primitive condition in the earliest angiosperm.

Answer Key to Chapter 31 Study Guide Questions

After You Have Read the Chapter

1. The Koonwarra angiosperm, like modern angiosperms, had small flowers without petals, borne in a spikelike inflorescence; single carpels with short stigmas and no styles; imperfect flowers with bracts at the base; and similar leaf venation. (p. 742-745)
2. The primitive angiosperm flower is thought to have resembled that of the Koonwarra angiosperm. They would be small and without petals borne in spikes. The single carpel ovaries would have short stigmas with no styles. The flowers would have clusters of bracts at their bases. (p. 744-745)
3. Angiosperms evolution is currently thought to have taken place somewhere in the semiarid regions of Western Gondwanaland in the uplands where it was cool and dry. Rapid diversification of angiosperms occurred in the lowland regions of the Tertiary Period. (p. 746)
4. The diversification of beetle-like insects into lepidopterans and hymenopterans accompanied the diversification of angiosperms. Angiosperms and insects seemed to have co-evolved. (p. 747)
5. Biochemical, chromosomal, and microscopic analysis along with RNA sequencing and DNA sequencing of nuclear and chloroplast genes can be used to verify the classification of a plant group. (p. 751)

6. Floras provide a list of all the plants in an area along with descriptions and keys. It provides a resource for studying plants in an area and an historical record of changes in the flora over time. (p. 748-749)
7. Flowers appear to be derived from a bisporangiate compound strobilus that evolved from fertile leaves in a seed fern ancestor. (p. 752)
8. The angiosperm ancestor is thought to be derived from a group of seed ferns that may also have produced the gnetophytes and cycadeoids at a slightly later time. (p. 753)
9. Comparison of the nucleotide sequences in the *rbc*L gene allow one to determine relatedness. The greater the differences in the sequences, the less closely related are the two groups. When groups share similar altered sequences, it is assumed the sequences were inherited from a common ancestor. (p. 753)
10. The monocots and dicots are thought to have diverged from a common angiosperm ancestora very soon aftert he divergence of the angiosperms from a common a seed fern ancestor. Evidence from cladistic analysis places the origin of the angiosperms at least 200 million years ago. Sequencing of the *rbc*L and 11 other chloroplast genes gives a similar date for the divergence of the monocots and dicots. Sequence analysis of nucelar genes gives a divergence date as long ago as 320 million years. (p. 754)

Constructing a Concept Map

Compare your concept map to the sample in the introduction to the Study Guide.

Fill-in-the-Blanks

1. trees
2. Cretaceous
3. cupule
4. single
5. gymnosperms
6. Pangea
7. leaf
8. cool, dry
9. gymnosperms
10. insects
11. nectaries
12. Cycadeoids
13. floras
14. flower
15. rbcL gene
16. monocots
17. *Ceratophyllum*
18. bisporangiate compound strobilus
19. cycadeoids and gnetophytes
20. molecular clock hypothesis

Putting Your Knowledge to Work

1. A. DNA sequence analysis of the *rbc*L gene
2. D, compound carpels
3. C. far eastern regions of Canada
4. B. bisporangiate strobili.

Doing Botany Yourself

The earliest flowers are all perfect with free floral parts and radial symmetry. They may or may not have sepals and petals. They have subtending bracts. The leaves have angiosperm-like venation, and the pollen may be gymnosperm-like. The ancestor would be expected to have a bisexual reproductive structure derived from a branching system as in the flower. The fertile organs and sterile organs would be free, not fused. The reproductive organs would be arranged in whorls or spiraled with accompanying bracts at the base of the reproductive structure. The leaves of the ancestor would have net venation, not dichotomous venation, with early signs of the veins that dead end as in angiosperms.

CHAPTER 32 POPULATION DYNAMICS AND COMMUNITY ECOLOGY

After You Have Read the Chapter

1. What are the major components of an ecosystem?
2. Contrast biotic potential to carrying capacity.
3. Explain why a food chain has a limited number of trophic levels.
4. Describe the trophic levels of an ecosystem.
5. What are two theories that explain why species diversity is important to an ecosystem's stability over time..
6. Compare an organism's habitat to its niche.
7. In what ways do abiotic factors shape ecosystems?
8. Besides competing for resources, in what ways do organisms interact in an ecosystem?
9. Describe the steps in the nitrogen cycle.
10. Compare primary and secondary succession.
11. What products of civilized life can contribute to an increased greenhouse effect?

Constructing a Concept Map

Using the explanation in the introduction to the Study Guide, construct a concept map of the main ideas in the chapter. Use the headings in the text as source material.

Fill-in-the-Blanks

1. A ________________________ is a group of individuals of a single species. (p. 760)
2. All of the populations of living organisms in a given area is a ________________________. (p. 761)
3. The communities in an area and their physical environment make up a (an) ________________. (p. 761)
4. When several individuals in several local populations are distributed over a wide area, the population may be a (an) ________________. (p. 760)
5. A population of sugar maples growing on the eastern slopes of Mt. Washington may be an ______________. (p. 761) if its members are physiologically and morphologically distinct from other sugar maple populations.
6. The soil type on which a plant is growing would be a (an) ____________ factor. (p. 761)
7. ________________ are plants adapted to conditions in which water stress rarely, if ever, occurs.
8. Bacteria and fungi play the role of ________________ in an ecosystem. (p. 762)
9. In a grassland ecosystem their is ______________biomass per acre in grass than in grazing animals. (p. 742)
10. ____________________ is the only factor that must be replaced from outside the earth's ecosystems. (p. 763)
11. The harsher the climate and poorer the soil the lower the ____________________ of an ecosystem leaving it more vulnerable to disturbances. (p. 764-765)
12. Ants that feed on the oil appendages of bleeding heart plant the seeds of the bleeding heart. These two species have a (an) ____________ relationship. (p. 766)
13. The availability of usable ____________________ is critical to plants for the production of proteins.
14. Nitrogen is added to soil by the decomposition of dead organisms and by______________________. (p. 769 that live in the soil and form symbiotic relationships with plants.
15. In the carbon cycle over 90% of the fixed carbon is released to the atmosphere by ___________. (p. 770)
16. Evaporation from bodies of water and transpiration from vegetation generates ____________________ that falls to earth as precipitation. (p. 762)

17. The first plants to establish themselves in a succession are the ___________________ species. (p. 772)
18. As succession continues, the vegetation reaches an equilibrium with stable plant associations called _________________. (p. 773)
19. The dumping of raw sewage and agricultural runoff into lakes can accelerate _________________. (p. 774)
20. In grasslands, chaparral, and forest ecosystems, ____________ ___________convert accumulated organic matter to mineral-rich ash which releases minerals to the ecosystem. (p. 775)
21. A rise in global temperature may be caused by an intensification of the __________________. (p. 776)
22. The ____________ _______levels in the atmosphere are steadily increasing due to the burning of fossil fuels and the deforestation of large areas of the earth. (p. 777)
23. Ozone in the upper atmosphere decreases the amount of _____________________ reaching the surface of the earth. (p. 778)
24. Sunlight converts sulfur and nitrogen compounds produced from the burning of fossil fuels into oxides that fall to earth as ____________________. (p. 778)
25. Industrial waste, pesticides, fertilizers, and sewage are polluting our __________________, both surface and underground supplies. (p. 779)

Putting Your Knowledge to Work

1. If a tree converts sunlight into 100,000 calories of starch, how many trophic levels in a single food chain can those calories support if at least 10 calories must be available to the next trophic level?
   A. ten
   B. five
   C. three
   D. four

2. The ecosystem with the greatest species diversity is
   A. tundra.
   B. deciduous temperate forest.
   C. open tropical ocean.
   D. the Taiga.

3. A species of plant is better adapted than any other species to its
   A. habitat.
   B. niche.
   C. biome.
   D. edaphic factors.

4. In a lake left by a retreating glacier the first autotrophic species to appear would be
   A. algae.
   B. floating plants.
   C. cattails or reeds.
   D. duckweed.

5. A possible plan to counteract the greenhouse effect of burning fossil fuels would be to
   A. cut down the vast expanses of tropical rain forest that support large populations of methane-producing termites.
   B. fertilize the oceans and large lakes to increase the populations of algae.
   C. build huge air conditioning units in desert regions to cool the atmosphere.
   D. reforest areas undergoing desertification such as North Africa.

Doing Botany Yourself

You are asked to construct a terrestrial ecosystem in a container that requires only light after an initial watering. What are the basic components you would need?

Answer Key to Chapter 32 Study Guide Questions

After You Have Read the Chapter

1. Ecosystems consist of populations of organisms interacting to form distinct communities adapted to a specific set of environmental conditions and physical characteristics of an area. The living organisms make up the biotic factors in the ecosystem while the environmental conditions and physical characteristics are the abiotic factors. (p. 761)
2. The number of offspring an individual can produce that can, under ideal conditions, live to reproduce is its biotic potential. The maximum number of individuals thatcan survive and reproduce in an ecosystem is its carrying capacity. (p. 762)
3. A food chain has a limited number of trophic levels because each time energy stored by the organisms at one trophic level is converted to energy after being consumed by organisms at the next trophic level, some energy is lost to the surrounding environment. This energy cannot be recaptured and used to run biotic systems. The loss from one level to the next can be as much as 90%. At some point the amount of energy required to utilize the energy is more than the energy store. This limits food chains to three or sometimes four levels of organisms. (p. 762-763)
4. The trophic levels of an ecosystem consist of the producers, primary consumers that eat the producers, secondary consumers that eat the primary consumers, maybe tertiary consumers that eat the secondary consumers, and the decomposers that consume all the rest when they die. (p. 762)
5. The ``rivet popper" and redundancy hypotheses are two theories of the importance of species diversity. The rivet popper hypothesis holds that each species in an ecosystem plays a small role in the total functioning of the system. The loss of each species weakens the system until, when enough species are lost, the ecosystem is destroyed. The redundancy hypothesis theorizes that most of the species in an ecosystem are superfluous. Only a few key species are required to maintain a functioning system. (p. 764-765)
6. The specific environmental conditions to which an organism is adapted is its habitat. The role that an organism plays in that habitat is its niche. A niche usually involves the ways in which the organism affects the survival of other organisms with which it comes in contact in the habitat. (p. 765)
7. Abiotic factors affect ecosystems by setting the limits of the physical factors in the environment. These limits become environmental stresses to which organisms must adapt to survive. They also determine the types of plant communities that can develop and thus the animals that can live in the ecosystem. (p. 766)
8. Organisms can interact in a variety of ways. Some organisms interact by helping one another survive (mutualism); pollinators ensure the reproduction of plants; pioneer species add organic matter to the soil, making it suitable for other plants. Plant roots release carbon dioxide which breaks down rock into soil; some plants shade others; plants produce chemicals that retard the growth of other plants or bacteria and fungi; plants alter their nutrient and chemical content to alter the feeding habits of other species; plants can produce insecticides; and plants can form symbiotic and parasitic relationships with other organisms. (p. 767)
9. Nitrogen compounds enter the nitrogen cycle through the roots of plants. These compounds are produced by the fixation of nitrogen into nitrates, nitrites, or ammonia through the action of free-living bacteria, symbiotic bacteria, or cyanobacteria. Other nitrogen is made available by the decomposition of organic matter released by dead organisms or in waste products. Nitrogen is released back to the atmosphere to be re-fixed by denitrifying bacteria, fire, and other means. Precipitation returns to the soil nitrogen compounds from the atmosphere where they accumulated due to industrial pollution, lightning, and the diffusion of ammonia from decomposition to the soil. (pp. 769)
10. Primary succession occurs in an environment where nutrients are in low supply. The succession works to increase the nutrient content of the ecosystem through the action of pioneer species and the accumulation of windblown soils. In secondary succession the soil base usually remains intact along with a substantial amount of nutrients. The disturbance before the secondary succession has disrupted some, but not all, of the interrelationship between the organisms and their environment in the area. (p. 772-774)

11. Massive forest fires, burning of fossil fuels, clearing of tropical rain forests, chlorofluorocarbons, and raising grazing animals can all lead to an increase in the greenhouse effect. (p. 775)

Constructing a Concept Map

Compare your concept map to the sample in the introduction to the Study Guide.

Fill-in-the-Blanks

1. population
2. biotic community
3. ecosystem
4. ecological race
5. ecotype
6. abiotic
7. Hydrophytes
8. decomposers
9. more
10. energy
11. species diversity
12. mutualistic
13. nitrogen
14. nitrogen-fixing bacteria
15. respiration
16. water vapor
17. pioneer
18. climax communities
19. eutrofication
20. fires
21. greenhouse effect
22. carbon dioxide
23. ultraviolet radiation
24. acid rain
25. water supplies

Putting Your Knowledge to Work

1. D., four
2. B., temperate deciduous forest
3. B. niche
4. A. algae
5. D. reforest areas undergoing desertification such as North Africa

Doing Botany Yourself

You will need to include decomposers (soil fungi and bacteria), producers (various types of plants), and consumers (herbivores and carnivores and/or insectivores). Soil microorganisms are associated with particular soil types that support particular groups of plants. Choosing compatible microbes and plants is the basic requirement for constructing a successful ecosystem. Choose animals whose size is compatible with your container and whose nutritional needs can be met by the producers the container can support. Some experimentation will be needed to establish a functioning system.

CHAPTER 33 BIOMES

After You Have Read the Chapter

1. Describe the interaction between climate and vegetation in the formation of the world's biomes.
2. How does permafrost affect the vegetation of the arctic tundra?
3. What types of abiotic factors challenge the survival of plants in the taiga?
4. How has the North America temperate deciduous forest biome changed since the arrival of European colonists?
5. Describe the features shared by the North American prairie and the African savannah?
6. What are the key factors that create a desert biome?
7. Describe the types of adaptations that have evolved in desert plants that aide in their survival.
8. Describe the role of fire in the maintenance of biomes.
9. How are tropical rain forests unique biomes?
10. Compare the forest biomes of the Northern Hemisphere.

Constructing a Concept Map

Using the explanation in the introduction to the Study Guide, construct a concept map of the main ideas in the chapter. Use the headings in the text as source material.

Fill-in-the-Blanks

1. Each of the earth's ____________________ can be recognized by its vegetation. (p. 783)
2. Much more rain will fall on the leeward side of a mountain range than in its ____________________. (p. 784)
3. The soil of the tundra is shallow and nutrient-poor and is underlaid by ____________ which prevents water drainage. (p. 785-786)
4. ____________ are absent from the tundra, although low-growing shrubs can be a dominant plant type in some regions. (p. 786)
5. Taiga is characterized by ____________________ forests. (p. 787)
6. The taiga biome has winter days of less than six hours of ____________________ and low temperatures of less than −50°C. (p. 787)
7. The floras of temperate forests of eastern North America are dominated by ____________________ tree species like maples and oaks. (p. 788-789)
8. Temperate deciduous forests can be found in North America, ____________, and Asia. (p. 788)
9. Fungal diseases have nearly eliminated ____________ trees and ____________ trees from eastern temperate deciduous forests in North America. (p. 789)
10. In temperate deciduous forests most wild flowers bloom in the early ____________ before the leafy canopy of the forest shades the forest floor. (p. 790)
11. Natural ____________ usually occur in the interiors of continents and along arid coastlines in temperate regions. (p. 790)
12. ____________________ are tropical grasslands with seasonal rains and scattered trees. (p. 791)
13. The American prairie has been converted into rangeland for cattle and cropland for growing ____________ and ____________. (p. 792)

14. ____________________ form in regions with low relative humidity between 20° and 30° latitude. (p. 792)
15. The _____________ in deserts fluctuates widely between day and night. (p. 793)
16. Many desert plants have adaptations to protect them from the effects of high temperature and low levels of available _____________. (p. 793)
17. ____________________ biomes have high concentrations of shrubs that are either evergreen or deciduous in the summer. (p. 795)
18. Chaparral dominates he Pacific coast of southern ______________________. (p. 796)
19. The mountain forests of the Western US are dominated by ____________________ species interspersed with large stands of deciduous trees like aspens. (p. 797)
20. Today much of the moisture available to the redwoods of the California and Oregon coast occurs in the form of ______________________. (p. 798)
21. Cones of some pine trees only release their seeds when exposed to _____________. (p. 799)
22. The greatest biological diversity in any biome occurs in the ______________________. (p. 800)
23. Broadleaf __________________ trees dominate the tropical rain forests along with many lianas and epiphytes. (p. 800)
24. The greatest threat to the survival of the tropical rain forest is tropical rain forests in our lifetime is ______________________________. (p. 802)

Putting Your Knowledge to Work

1. An ecologist is trying to determine the biome in which his laboratory is located. What should he study in order to make this determination?
   A. the native vegetation
   B. the dominate soil type in the region
   C. the average annual rainfall
   D. the herbivores in the area

2. Pine trees growing in the taiga are not likely to
   A. be tolerant of ice and snow.
   B. tolerate short summer seasons.
   C. have cones that release seeds in response to fire.
   D. grow in acid soils.

3. In the Rocky Mountains of western North America what biome probably would not occur at any elevation?
   A. taiga
   B. alpine tundra
   C. savanna
   D. mountain forest

4. You want to build a ranch in California. You do not want to have to irrigate to produce enough forage for your cattle. Where should you locate the ranch?
   A. in the southeastern interior
   B. on the southern coast
   C. on the western slopes of the northern mountains
   D. on the eastern slopes of the southern mountains

Doing Botany Yourself

You are teaching a course in plant ecology. You have a limited amount of money. You want to expose your students to as many biomes as possible in the shortest time over the shortest distance. Develop a plan to meet these goals.

Answer Key to Chapter 33 Study Guide Questions

After You Have Read the Chapter

1. Climate determines the amount and pattern of precipitation, available light, length of the growing season, temperature ranges, and average daily temperature. All of these factors determine the types of plants that can survive in the area. The types of plants then determine the fauna that will be able to establish itself in the region. (p. 784)
2. Permafrost prevents the growth or deep-rooted plants like trees limiting arctic tundra vegetation to grasses, sedges, mosses, and lichens. The growth of large plants is limited because the permafrost creates anaerobic conditions with water-logged, nutrient-poor soils during the growing seasons. (p. 785)
3. The taiga biome is a northern coniferous forest adjacent to the arctic tundra. Plants growing there must contend with permafrost to within a meter of the ground surface, acidic and nutrient-poor soils, a very short growing season, and a yearly temperature range is from 27 degrees C in the summer to -50 degrees C in the winter. (p. 787)
4. Most of thew forests have been cut and replaced by farmland, cities, or second growth forests. lWidespread and abundant species such as American elm and the American chestnut have been virtually eliminated by fungal diseases. (p. 788-789)
5. In both the prairie and savannas the climax vegetation consists mainly of grasses. Herbaceous perennials with bulbs, rhizomes, and other underground storage structures are also abundant. (p. 790-791)
6. Deserts occur between 20° and 30° latitude. They occur in the interiors of continents and on the coasts where conditions limit rainfall and relative humidity, creating drying conditions and fluctuating daily temperatures. (p. 790)
7. In grassland biomes fires are frequent and are essential for maintaining the grassland species by freeing up nutrients and removing competing species. In Mediterranean scrub biomes fire are frequent and can be required for seed germination. Fires in mountain forest biomes conifer seeds may only be released and/or germinate in response to fire. (p. 796-799)
8. Desert plants are adapted to withstand high temperatures, low rainfall, and low relative humidity by having adaptations such as the following: CAM photosynthesis, C_4 photosynthesis, thick cuticles, few stomata, water-storage tissue, reduced or leathery leaves, deciduous leaves, no leaves, deep taproots, widespread shallow roots, and bulbs. (p. 793)
9. Tropical rain forests have a high level of species diversity that exceeds all other biomes. The biome has existed for over 200 million years intact. Temperature ranges fluctuate only a few degrees daily and year-round. Rainfall can be 200 to 400 cm per year and relative humidity exceeds 80%. The vegetation is nearly all broadleaf evergreen trees with no conifers, some lianas, and some epiphytes. (p. 800-802)
10. The Northern Hemisphere has the taiga, temperate deciduous forests, and mountain forests. The forests vary mainly in the frequency and types of conifers present in the forests which is a function of climate and soil conditions. Taiga is completely conifer forest with the coldest temperatures and the most nutrient-poor soils. Mountain forests are also dominated by conifers with the type and percentage varying from region to region. The climate is generally cooler than that at lower altitudes in the region and may be very moist. Temperate forests have hardwoods predominating with more conifers toward the northern and southern edges of the biome. Temperatures vary seasonally but are generally more moderate than in the taiga. The available moisture varies but is moderate. (p. 784-790)

Constructing a Concept Map

Compare your concept map to the sample in the introduction to the Study Guide.

Fill-in-the-Blanks

1. biomes
2. rain shadow
3. permafrost
4. Trees
5. conifers
6. daylight
7. deciduous
8. Europe
9. chestnut, elm
10. spring
11. grasslands
12. Savannas
13. corn, wheat
14. Deserts
15. temperature
16. moisture
17. Mediterranean scrub
18. California
19. conifer
20. fog
21. fire
22. tropical rain forests
23. evergreen
24. human activities

Putting Your Knowledge to Work

1. A., the native vegetation
2. C. have cones that release seeds in response to fire
3. C. savanna
4. C. on the western slopes of the northern mountains

Doing Botany Yourself

Vegetation changes with latitude and altitude. When ascending mountains located at a specific latitude one will see changes in vegetation equivalent to changes encountered by moving to a more northern latitude within the same longitude.

You can take advantage of this to quickly expose your students to a variety of biomes in a limited space and time.

Take your students to the Blue Ridge or Smoky Mountains of Tennessee. These mountains have peaks above 5,000 feet. The change in altitude as you climb the mountains will take students from the southern temperate deciduous biome of Tennessee through more northern temperate deciduous biomes to the taiga and in a few places alpine tundra at the mountain peaks. It is the equivalent of a trip from Tennessee to near the Arctic Circle.

Have the students take species counts along transects at every 1000 feet in elevation. Have them compare the species composition to biomes characteristic of areas lying to the north. Let them develop a relationship between increase in latitude and increasing northern latitude using this information.